SpringerBriefs in Applied Sciences and Technology

Forensic and Medical Bioinformatics

Series editors

Amit Kumar, Hyderabad, India
Allam Appa Rao, Hyderabad, India

More information about this series at http://www.springer.com/series/11910

Naresh Babu Muppalaneni
Vinit Kumar Gunjan
Editors

Computational Intelligence in Medical Informatics

 Springer

Editors
Naresh Babu Muppalaneni
C.R. Rao Advanced Institute
   of Mathematics, Statistics
   and Computer Science
Hyderabad
India

Vinit Kumar Gunjan
Annamacharya Institute of Technology
   and Sciences
Kadapa
India

ISSN 2191-530X   ISSN 2191-5318   (electronic)
ISBN 978-981-287-259-3   ISBN 978-981-287-260-9   (eBook)
DOI 10.1007/978-981-287-260-9

Library of Congress Control Number: 2014953255

Springer Singapore Heidelberg New York Dordrecht London

Printed on acid-free paper

Springer is part of Springer Science+Business Media (www.springer.com)

Foreword

This book is an enthusiastic contribution of the best research work in the field of bioinformatics, biotechnology, and allied domains to the International Conference on Computational Intelligence: Health and Disease (CIHD 2014) to be held at Visakhapatnam, India during December 27–28, 2014. The main objective of this conference is to create an environment for (1) cross-disseminating state-of-the-art knowledge to CI researchers, doctors and computational biologists; (2) creating a common substrate of knowledge that both CI people, doctors and computational biologists can understand; (3) stimulating the development of specialized CI techniques, keeping in mind the application to computational biology; (4) fostering new collaborations among scientists having similar or complementary backgrounds.

Yet another element is provided by many interesting historical data on diabetes and an abundance of colorful illustrations. On top of that, there are innumerable historical vignettes that interweave computer science and biology in a very appealing way.

Although the emphasis of this work is on diabetes and other diseases, it contains much that will be of interest to those outside this field and to students of Biotechnology, Bioinformatics, Chemistry and Computer Science—indeed to anyone with a fascination for the world of molecules. The authors have selected a good number of prominent molecules as the key subjects of their essays. Although these represent only a small sample of the world of biologically related molecules and their impact on our health, they amply illustrate the importance of this field of science to humankind and the way in which the field has evolved.

I think that the contributors can be confident that there will be many grateful readers who will have gained a broader perspective of the disciplines of diabetes and their remedies as a result of their efforts.

Hyderabad, India

Naresh Babu Muppalaneni
Vinit Kumar Gunjan

Preface

This volume contains a selection of the best contributions delivered at the International Conference on Computational Intelligence: Health and Disease (CIHD 2014) held at Visakhapatnam, India during December 27–28, 2014. This conference is organized by Institute of Bioinformatics and Computational Biology (IBCB), Visakhapatnam, India jointly with Andhra University and JNTU Kakinada.

The IBCB is a research organization established in Visakhapatnam, India. It is a community of scholars devoted to the understanding of mysteries that remain in the catalogue of human genes through intellectual inquiry. The Institute encourages and supports curiosity-driven research in the fields of Bioinformatics and Computational Biology. The institute nurtures speculative thinking that produces advances in knowledge that change the way we understand the world. It provides for the mentoring of scholars, and ensures the freedom to undertake research that will make significant contributions in any of the broad range of fields in Bioinformatics and Computational Biology.

CIHD 2014 is aimed to bring together computer professionals, doctors, academicians, and researchers to share their experience and expertise in Computational Intelligence. The goal of the conference is to provide computer science professionals, engineers, medical doctors, bioinformatics researchers, and other interdisciplinary researchers a common platform to explore research opportunities.

A rigorous peer-review selection process was applied to ultimately select the papers included in the program of the conference. This volume collects the best contributions presented at the conference.

The success of this conference is to be credited to the contribution of many people. In the first place, we would like to thank Prof. Allam Appa Rao, Director, C.R. Rao AIMSCS, who motivated and guided us in making this conference a grand success. Our sincere thanks to Dr. Amit Kumar, Editor for Springer Briefs in Applied Sciences and Technology, who helped us in bringing this series. Moreover, special thanks are due to the Program Committee members and reviewers for their commitment to the task of providing high-quality reviews. We thank Prof. B.M. Hegde (Padma Bhushan Awardee, Cardiologist and Former Vice Chancellor, Manipal University) who delivered the keynote address. Last but not least, we would thank the speakers Grady

Hanrahan (California Lutheran University, USA), Jayaram B. (Coordinator, Super-computing Facility for Bioinformatics and Computational Biology, IIT Delhi), Jeyakanthan J. (Professor and Head, Structural Biology and Biocomputing Lab, Alagappa University), Nita Parekh (International Institute of Information Technology Hyderabad (IIIT-H), Hyderabad, India), Pinnamaneni Bhanu Prasad (Advisor, Kelenn Technology, France), Rajasekaran E. (Dhanalakshmi-Srinivasan Institute of Technology, Tiruchirappalli), and Sridhar G.R. (Endocrine and Diabetes Centre, Krishnanagar Visakhapatnam, India).

December 2014   Naresh Babu Muppalaneni
                                                                             Vinit Kumar Gunjan

Committees

International Conference on Computational Intelligence: Health and Disease (CIHD 2014)

27–28 December 2014
Visakhapatnam, India

The International Conference on Computational Intelligence: Health and Disease (CIHD 2014) held at Visakhapatnam, India during December 27–28, 2014 is organized by Institute of Bioinformatics and Computational Biology (IBCB), Visakhapatnam, India jointly with Andhra University and JNTU Kakinada.

General Chair

Dr. Allam Appa Rao, SDPS Fellow, Director, C.R. Rao AIMSCS, UoH, Hyderabad, India

Conference Secretary

Dr. P. Sateesh, Associate Professor, MVGR College of Engineering

Organizing Committee

Dr. Ch. Divakar, Secretary, IBCB

Prof. P.V. Nageswara Rao, Head of the Department, Department of CSE, GITAM University

Prof. P.V. Lakshmi, Head of the Department, Department of IT, GITAM University

Prof. P. Krishna Subba Rao, Professor, Department of CSE, GVP College of Engineering (Autonomous)

Dr. G. Satyavani, Assistant Professor, IIIT Allahabad

Dr. Akula Siva Prasad, Lecturer, Dr VS Krishna College

Shri. Kunjam Nageswara Rao, Assistant Professor, AU College of Engineering

Shri. D. Dharmayya, Associate Professor, Vignan Institute of Information Technology

Shri. T.M.N. Vamsi, Associate Professor, GVP PG College

Advisory Committee

Prof. P.S. Avadhani, Professor, Department of CS and SE, AU College of Engineering

Prof. P. Srinivasa Rao, Head of the Department, Department of CS and SE, AU College of Engineering

Dr. Raghu Korrapati, Professor, Walden University, USA

Prof. Ch. Satyanarayana, Professor, Department of CSE, JNTU Kakinada

Prof. C.P.V.N.J. Mohan Rao, Professor and Principal, Avanthi Institute of Engineering and Technology

Dr. Anirban Banerjee, Assistant Professor, IISER Kolkata

Dr. Raghunath Reddy Burri, Scientist, GVK Bio Hyderabad

Dr. L. Sumalatha, Professor and Head, Department of CSE, JNTU Kakinada

Dr. D. Suryanarayana, Principal, Vishnu Institute Technology, Bhimavaram

Dr. A. Yesu Babu, Professor and Head, Department of CSE, Sir C.R. Reddy College of Engineering, Eluru

Dr. T.K. Rama Krishna, Principal, Sri Sai Aditya Institute of Science and Technology

Finance Committee

Shri. B. Poorna Satyanarayana, Professor, Department of CSE, Chaitanya Engineering College

Dr. T. Uma Devi, GITAM University

Dr. R. Bramarambha, Associate Professor, Department of IT, GITAM University

Smt. P. Lakshmi Jagadamba, Associate Professor, GVP

Smt. Amita Kasyap, Women Scientist, C.R. Rao AIMSCS

Publication Committee

Dr. Amit Kumar, Publication Chair, Director, BDRC
Dr. Kudipudi Srinivas, Co-chair, Professor, V.R. Siddhartha Engineering College
Dr. G. Lavanya Devi, Assistant Professor, Department of CS and SE, AU College of Engineering
Dr. P. Sateesh, Associate Professor, MVGR College of Engineering
Dr. A. Chandra Sekhar, Principal, Sankethika Institute of Technology
Dr. K. Karthika Pavani, Professor, RVR and JC College of Engineering

Website Committee

Dr. N.G.K. Murthy, Professor of CSE, GVIT Bhimavaram
Dr. Suresh Babu Mudunuri, Professor of CSE, GVIT Bhimavaram
Shri. Y. Ramesh Kumar, Head of the Department, CSE, Avanthi Institute of Engineering and Technology

Financing Institutions
Department of Science and Technology, Government of India

Contents

Analysis of the Structural Details of DsrO Protein from *Allochromatium vinosum* to Identify the Role of the Protein in the Redox Transport Process Through the *dsr* Operon

Semanti Ghosh and Angshuman Bagchi

Abstract Sulfur oxidation is one of the oldest known redox processes in our environment mediated by phylogenetically diverse sets of microorganisms. The sulfur oxidation process is mediated mainly by *dsr* operon which is basically involved in the balancing and utilization of environmental sulfur compounds. DsrMKJOP complex from the *dsr* operon is the central player of this operon. DsrO is a periplasmic protein which binds FeS clusters responsible for electron transfer to DsrP protein from the *dsr* operon. DsrP protein is known to be involved in electron transfer to DsrM protein. DsrM protein would then donate the electrons to DsrK protein, the catalytic subunit of this complex. In the present work, we tried to analyze the role of DsrO protein of the *dsr* operon from the ecologically and industrially important organism *Allochromatium vinosum*. There are no previous reports that deal with the structural details of the DsrO protein. We predicted the structure of the DsrO protein obtained by homology modeling. The structure of the modeled protein was then docked with various sulfur anion ligands to understand the molecular mechanism of the transportation process of sulfur anion ligands by this DsrMKJOP complex. This study may therefore be considered as a first report of its kind that would therefore enlighten the pathway for analysis of the biochemical mechanism of sulfur oxidation reaction cycle by *dsr* operon.

Keywords Sulfur oxidation · Ecological importance · *dsr* operon · Dsro protein · Homology modeling · Molecular docking

S. Ghosh · A. Bagchi (✉)
Department of Biochemistry and Biophysics, University of Kalyani,
Kalyani 741235, Nadia, India
e-mail: angshuman_bagchi@yahoo.com

© The Author(s) 2015
N.B. Muppalaneni and V.K. Gunjan (eds.), *Computational Intelligence in Medical Informatics*, Forensic and Medical Bioinformatics,
DOI 10.1007/978-981-287-260-9_1

1

1 Introduction

Sulfur oxidation reaction cycle is one of the important biogeochemical cycles in the world. Sulfur has a wide range of oxidation states viz., +6 to −2. This makes the element capable of taking part in a number of different biological processes. Sulfur-based chemo or photolithotrophy is one of such processes involving the transfer of electrons from reduced sulfur compounds like sulfite, thiosulfate, elemental sulfur, etc. The sulfur oxidation process is mediated by a diverse set of microorganisms. Only little is known about the molecular mechanisms of this sulfur oxidation process in these microorganisms. One of the sulfur oxidizers is *Allochromatium vinosum* (*A. vinosum*), a dominant member of purple sulfur bacteria. This bacterium uses reduced sulfur compounds as electron donor for anoxygenic photosynthesis [1]. Recent studies with *A. vinosum* revealed that a multiple gene cluster comprising genes *dsrA*, *dsrB*, *dsrE*, *dsrF*, *dsrH*, *dsrC*, *dsrM*, *dsrK*, *dsrL*, *dsrJ*, *dsrO*, *dsrP*, *dsrN*, *dsrS*, and *dsrR* is involved in the sulfur oxidation process [2]. The organism *A. vinosum* has a wide range of applications in different industrial processes like waste remediation and removal of toxic compounds, e.g., odorous sulfur compounds like sulfide and explosives and production of industrially relevant organo-chemicals such as vitamins, bio-polyesters, and biohydrogen [1]. It is well known that *A. vinosum* uses the DsrMKJOP protein complex to carry out the sulfur oxidation process [3]. DsrJ, a periplasmic protein, may be involved in the oxidation of a putative sulfur substrate in the periplasm and the released electrons would be transported across the membrane via the other components, viz., DsrO, DsrP, DsrM, DsrK successively, of the DsrMKJOP complex [3]. DsrO is a periplasmic iron-sulfur protein and is known to be involved in electron transfer process [2–4]. It is not yet known which amino acids of DsrO are involved in the electron transport or interact with sulfur anions. So, in this work, we have attempted to characterize DsrO protein at the structural level. We have predicted the putative active site geometry of the DsrO protein. In order to predict the molecular mechanism of the electron transport through DsrO protein we have docked the different sulfur anions present in the environment with DsrO protein. Till date there are no reports that deal with the analyses of the detailed structural information of DsrO as well as binding of sulfur anions with this protein to predict the mechanism of electron transport. This work is therefore the first of its kind. Since there are no previous reports regarding the molecular and structural details of DsrO protein, our work would therefore be important to analyze the biochemical mechanism of sulfur oxidation process by this ecologically and industrially important microbial species.

DsrO   PLMCQHCEHPPCVDVCPTGASFKRADGIVMVDRHLCIGCRYCMMACPYKARSFIHQPTTG 187

2VPW   PEQCLHCENPPCVPVCPTGASYQTKDGLVLVDPKKCIACGACIAACPYDAR-YLHP---------- 109

DsrO   QLTAVPRGKGCVESCNLCVHRRDNGEESTACVDAC 221

2VPW   ------------------AGYVSKCTFCAHRLEKG-KVPACVETC 135

Fig. 1 Sequence alignment of the DsrO functional domain with PsrABC (PDB code: 2VPW) as template

2 Materials and Methods

2.1 Sequence Analysis and Homology Modeling of DsrO Protein

The amino acid sequence of DsrO protein from *A. vinosum* was obtained from Entrez database (Accession No. YP_003443232). To identify the presence of domains in DsrO protein, the amino acid sequence of DsrO protein was used to search Pfam [5]. The Pfam results have shown the presence of the domain 4Fe-4S dis-cluster spanning amino acid residues 125–224. This protein superfamily includes proteins that bind to iron-sulfur clusters. The amino acid sequence of DsrO protein has been used to search Brookhaven Protein Data Bank (PBD) [6] using the software BLAST [7] for finding suitable template for homology modeling. The BLAST search result of DsrO has revealed it to have 45 % identity with X-ray crystal structure of Polysulfide (PsrABC) from *Thermus thermophilus* (*T. thermophilus*) (PDB code: 2VPW) [8, 9]. The sequence alignment is shown in Fig. 1.

The protein has been modeled using the corresponding crystal structure as template. Homology modeling has been performed using the software suite DS Modeling [MSI/ and Accelrys, San Diego, CA, USA].

The modeled structure has been superimposed on the crystal template without altering the coordinate system of atomic positions in the template. The root-mean-square deviation (r.m.s.d.) for the superimposition is 0.906 Å for DsrO with its template (Fig. 2).

The model is then subjected to energy minimization using the CHARMm force field [10] fixing the backbone of the modeled protein to ensure proper interactions until the structure reached the final derivative of 0.001 kcal/mol.

2.2 Validation of the Model

The stereo-chemical qualities of the three-dimensional model obtained after energy minimization are checked by VERIFY3D [11] and PROCHECK [12] using Structural Analysis and Verification Server (SAVes). Regarding the main chain

Fig. 2 Superimposition of
the α-carbon backbone of
DsrO on 2VPW (B). DsrO is
presented in *red* and 2VPW in
green

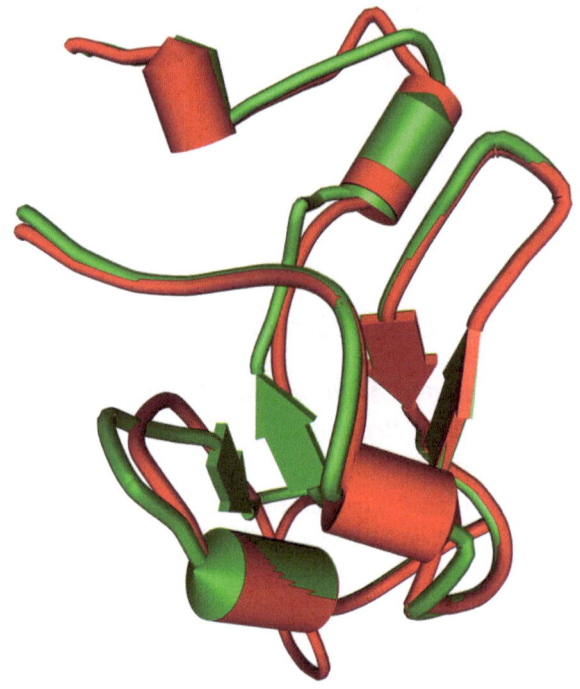

properties of the modeled protein neither considerable bad contacts nor C_α tetra-
hedron distortion nor hydrogen bond energy problems have been found. Moreover,
the average G factor, the measure of the normality degree of the protein properties,
has been found to be −0.10, which is inside the permitted values for homology
models. Furthermore, no distortions of the side chain torsion angles are found. The
Ramachandran plot [13] has been drawn. No residues are found to be present in the
disallowed regions of the Ramachandran plot. The residue profile of the model has
been checked by VERIFY3D and it indicates a good model quality.

2.3 Molecular Docking of DsrO with Sulfur Anions

In order to identify the active site residues in DsrO protein, InterPro database was
searched using the amino acid sequence of DsrO protein. The search result revealed
that DsrO has similar domain architecture to InterPro entry IPR017900. It has been
observed that four cysteine residues at amino acid positions 162, 165, 168, and 172
[IPR017900] are predicted to be involved in binding of sulfur anions. These active
site amino acid residues are used for docking purposes. To study the interactions
between DsrO and various sulfur anions that take part in sulfur oxidation process,
the model of DsrO protein has been docked with sulfite (SO_3^{2-}), sulfate (SO_4^{2-}),

and thiosulfate ($S_2O_3^{2-}$) using the program GOLD [14]. The docked complexes that have yielded the best GoldScore and ChemScore are selected and analyzed to study the interactions.

2.4 Molecular Mechanics Simulations of DsrO-Sulfur Anion Complexes

The docked complexes of DsrO with different sulfur anions have been subjected to molecular mechanics simulations. During simulation, the backbone of the DsrO protein has been kept fixed. The simulations have been performed in explicit solvent with a di-electric constant of 80. For the simulations, 2,000 cycles of steepest descent followed by 1,000 cycles of conjugate gradient methods have been used with CHARMm [10] force fields. The final coordinates of the docked complexes have been saved for analyses.

2.5 Calculations for Protein–Ligand Interactions

The stereo-chemical qualities of the energy minimized docked complexes are checked by PROCHECK and VERIFY3D. The interactions between the sulfur anions and DsrO protein are then analyzed by DISCOVERY STUDIO software suite.

3 Results and Discussion

3.1 Description of the Three-Dimensional Structure of DsrO from Allochromatium vinosum

The DsrO protein from *A. vinosum* has a functional domain consisting of 95 amino acids. The modeled structure is similar to X-ray crystal structure of Polysulfide Reductase (PsrABC) from *T. thermophilus* strain HB27 (PDB code: 2VPW, Chain B). The predicted structure of DsrO protein from *A. vinosum* consists of a mixture of helices, coils, and sheets (Fig. 3). The functional domain starts with a coil (amino acid residues 127–136) followed by a helix (amino acid residues 137–140). Then, there are two beta sheets (amino acid residues 147–149 and 155–157) and three helices (amino acid residues 167–171, 205–210, 216–219) joined by nine coil regions (amino acid residues 142–146, 150, 158, 162–166, 172–204, 211, 214–215, 220–221).

Fig. 3 Three-dimensional
model of DsrO protein. The
helices are shown in *red*. The
strands are presented in
yellow. The rest of the part is
coil region shown in *green*

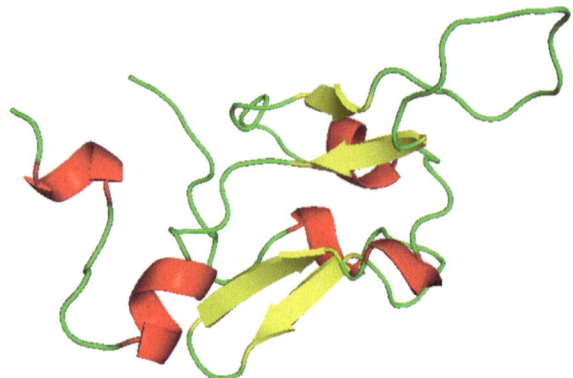

3.2 Putative Active Site Geometry of DsrO
from Allochromatium vinosum

DsrO belongs to the 4Fe-4S ferredoxin, iron-sulfur binding family of proteins. Proteins belonging to this class are involved in transport of electrons in a range of metabolic reactions; they fall into several subgroups according to the nature of their iron-sulfur cluster(s) [15, 16]. The various Fe-S sites found in many electron transfer proteins (e.g., ferredoxins) are also found in many enzymes, where these centers are involved in intra or inter-protein electron transfer processes. For example, sulfite reductase contains a siroheme and Fe_4S_4 center, which are strongly coupled and involved in the six electron reduction of SO_3^{2-} to H_2S [17]. In most ferredoxins, the protein ligands are cysteines, which provide four thiolate donors to the 1Fe, 2Fe, or 4Fe centers [17]. IPR017900 entry represents at least one conserved domain, including four cysteines residues that bind to a 4Fe-4S center. Alignment of the amino acid sequence of DsrO from *A. vinosum* has revealed that DsrO also has four cysteine residues at the identical positions viz., Cys162, Cys165, Cys168, Cys172. During molecular docking in GOLD program Cys168 was used as the binding site of DsrO protein with the sulfur anions as it covers all the required flexible side chains for docking. The flexible side chains of the amino acid residues which are mainly involved in the interactions with sulfur anions are listed in Table 1. This positioning of the amino acid residues is known to play an important role in ligand binding described in the next section.

Table 1 List of amino acids of DsrO protein involved in protein–ligand interactions

DsrO	Cys130, Cys138, Val139, Asp140, Val141, Cys142, Thr144, Ser147, Phe148, Val155, Met156, Val157, Asp158, Cys162, Ile163, Cys165, Arg166, Tyr167, Cys168, Met169, Met170, Cys172, Tyr174, Leu175, Ser178, Phe179, Ile180, Cys197, Val198, Glu199, Ser200

Table 2 Comparative values of GOLD program of protein–ligand complexes

Protein–ligand complex name	GoldScore fitness	ChemScore DG	ChemScore H-bond weighted
DsrO-thiosulfate	36.9534	−16.0675	−10.5875
DsrO-sulfate	30.7964	−13.9717	−8.4917
DsrO-sulfite	26.9615	−17.7079	−12.2279

3.3 GOLD Docking Scores for Protein–Ligand Interactions

The energetics of DsrO-sulfur anion interactions were calculated by considering the parameters of GOLD program as listed in Table 2. We have used thiosulfate, sulfite, and sulfate anions as the potential sulfur substrates since the actual sulfur anion ligand is not yet known clearly.

From the GoldScore fitness values among three complexes it may be safely concluded that DsrO protein binds strongly with thiosulfate. It was previously shown that sulfide/thiosulfate grown cells show high expression of DsrO protein [4]. Now, from these docking complexes it is confirmed that thiosulfate is the most suitable ligand among the different sulfur anions.

3.4 Interactions of DsrO with Sulfur Anions

The binding energy values associated with the docked complexes of DsrO-thiosulfate, DsrO-sulfate, and DsrO-sulfite are presented in Table 3.

It has also been observed that more number of amino acid residues of DsrO protein is involved in binding with thiosulfate as compared to other sulfur anions (Fig. 4). The amino acid residues Cys142, Gly164, Cys165, Tyr167, and Cys168 from DsrO protein are involved in binding with thiosulfate. On the other hand, for DsrO-sulfite complex the amino acids Gly164, Cys165, Arg166 are involved in binding, whereas for DsrO-sulfate complex, the amino acid residues Cys162 and Gly164 are involved in binding (Figs. 5 and 6). From this analysis it is apparent that DsrO protein binds strongly with thiosulfate as greater number of amino acid residues are involved in interactions. Thiosulfate being the largest of the three sulfur

Table 3 Binding energy calculations of protein–ligand complexes

Docked complex	Total binding energy of the protein–ligand complex	Energy of the complex	Entropy
DsrO-thiosulfate	−0.0975	−3717.26643	17.03740
DsrO-sulfite	0.55491	−3671.98443	17.01440
DsrO-sulfate	0.31514	−3655.62403	16.59680

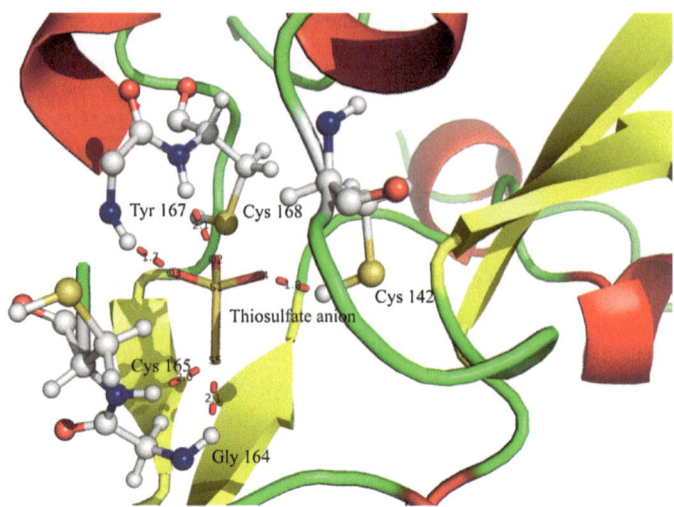

Fig. 4 Interactions of the DsrO protein with thiosulfate. Cys142, Gly164, Cys165, Tyr167, Cys168 amino acids from DsrO protein are involved in binding with thiosulfate

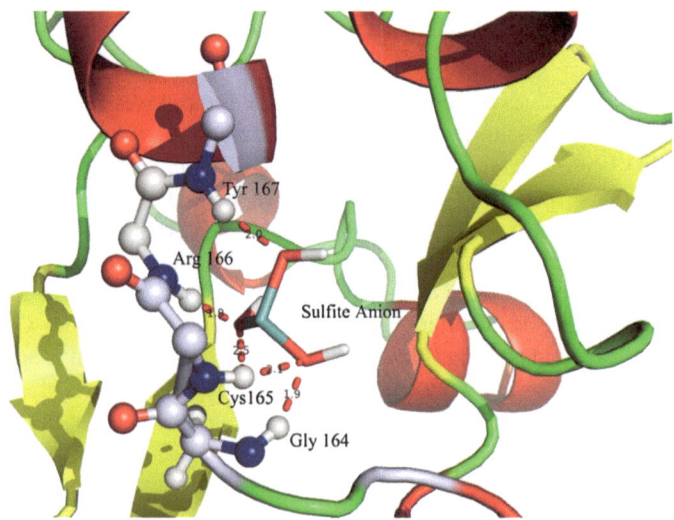

Fig. 5 DsrO-sulfite complex; the amino acids Gly164, Cys165, Arg166 are involved in binding

anion ligands used has the maximum chance of interactions with the protein. On the other hand in sulfate, the sulfur atom has the highest oxidation state of +6. Hence, sulfate has no ability to be oxidized and is not used in the oxidation process.

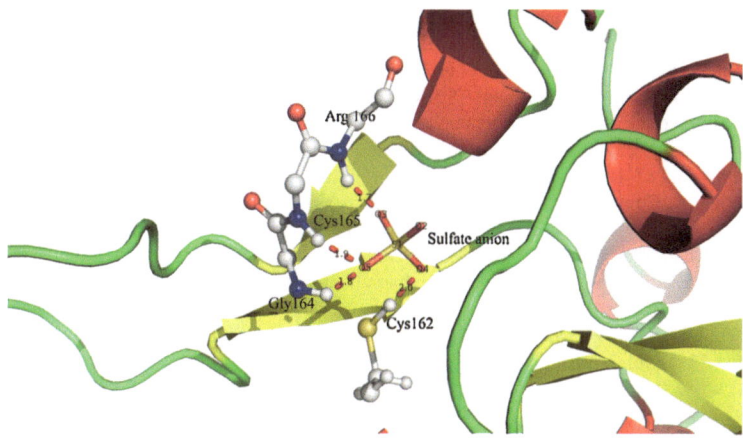

Fig. 6 DsrO-sulfate complex; the amino acid residues Cys162 and Gly164 are involved in binding

4 Conclusions

In this study we elucidate the structural basis of the involvement of DsrO protein from *A. vinosum* in electron transport during the oxidation of sulfur compounds. We built the three-dimensional structure DsrO protein using comparative modeling technique. The dockings of sulfur anions with DsrO allowed us to identify the details of their mode of interactions. We identified the amino acid residues from the DsrO protein that are involved in the binding of DsrO protein with sulfur anions. Results from this study will be important for understanding the pathway of electron transport via this protein in the global sulfur cycle. This homology model of DsrO provides a rational framework for designing experiments aimed at determining the contribution of amino acid residues responsible for electron transport via *dsr* operon.

Acknowledgments Ms. Semanti Ghosh is thankful to the University of Kalyani, Govt. of West Bengal, India, and UGC for the financial support. We would like to thank Bioinformatics Infrastructure Facility and the DST-PURSE program 2012–2015 going on in the department of Biochemistry and Biophysics, University of Kalyani for the support.

References

1. Weissgerber T, Zigann R, Bruce D, Chang Y, Detter JC, Han C, Hauser L, Jeffrie CD, Land M, Munk AC, Tapia R, Dahl C (2011) Complete genome sequence of *Allochromatium vinosum* DSM 180^T. Stand Genomic Sci 5:311–330
2. Grein F (2010) Biochemical, biophysical and functional analysis of the DsrMKJOP transmembrane complex from *Allochromatium vinosum*. PhD. thesis, Rhenish Friedrich Wilhelm University, Germany

3. Grein F, Pereira IAC, Dahl C (2010) Biochemical characterization of individual components of the *Allochromatium vinosum* DsrMKJOP transmembrane complex aids understanding of complex function in vivo. J Bacteriol 192(24):6369–6377
4. Dahl C, Engels S, Pott-Sperling AS, Schulte A, Sander J, Lübbe Y, Deuster O, Brune DC (2005) Novel genes of the dsr gene cluster and evidence for close interaction of Dsr proteins during sulfur oxidation in the phototrophic sulfur bacterium *Allochromatium vinosum*. J Bacteriol 187(4):1392–1404
5. Bateman A, Coin L, Durbin R, Finn RD, Hollich V, Griffiths-Jones S, Khanna A, Marshall M, Moxon S, Sonnhammer EL, Studholme DJ, Yeats C, Eddy SR (2004) The Pfam protein families database. Nucleic Acids Res 32(Database issue):D138-41
6. Berman HM (2008) The protein data bank: a historical perspective. Acta Crystallogr A 64:88–95
7. Altschul SF, Gish W, Miller W, Myers EW, Lipman DJ (1990) Basic local alignment search tool. J Mol Biol 215(3):403–410
8. http: www.rcsb.org/pdb/explore/explore.do?structureId=2VPW
9. Jormakka M, Yokoyama K, Yano T, Tamakoshi M, Akimoto S, Shimamura T, Curmi P, Iwata S (2008) Molecular mechanism of energy conservation in polysulfide respiration. Nat Struct Mol Biol 15(7):730–737
10. Brooks BR, Bruccoleri RE, Olafson BD, States DJ, Swaminathan S, Karplus M (1983) CHARMM: a program for macromolecular energy minimization and dynamics calculations. J Comp Chem 4:187–217
11. Lüthy R, Bowie JU, Eisenberg D (1992) Assessment of protein models with three-dimensional profiles. Nature 356(6364):83–85
12. Laskowski RA, McArthur MW, Moss DS, Thornton JM (1993) PROCHECK: a program to check the stereochemical quality of protein structures. J Appl Crystallogr 26:283–291
13. Ramachandran GN, Ramakrishnan C, Sasisekharan V (1963) Stereochemistry of polypeptide chain configurations. J Mol Biol 7:95–99
14. Jones G, Willett P, Glen RC, Leach AR, Taylor R (1997) Development and validation of a genetic algorithm for flexible docking. J Mol Biol 267(3):727–748
15. George DG, Hunt LT, Yeh LS, Barker WC (1985) New perspectives on bacterial ferredoxin evolution. J Mol Evol 22(1):20–31
16. Otaka E, Ooi T (1989) Examination of protein sequence homologies: V. New perspectives on evolution between bacterial and chloroplast-type ferredoxins inferred from sequence evidence. J Mol Evol 29(3):246–254
17. Bertini I, Gray HB, Lippard SJ, Valentine JS (1994) Bioinorganic chemistry. In: Stiefel EI, George GN (eds) Chapter 7: Ferredoxins, hydrogenases, and nitrogenases: metal-sulfide proteins. University Science Books, Mill Valley. ISBN:0-935702-57-1

LCMV Interaction Changes with T192M Mutation in Alpha-Dystroglycan

Simanti Bhattacharya, Sanchari Bhattacharjee,
Prosun Kumar Biswas, Amit Das, Rakhi Dasgupta
and Angshuman Bagchi

Abstract Limb girdle muscular dystrophy (OMIM: 613818) is a severe disease in humans, which broadly affects brain development. The disease is caused by T192M mutation in the protein alpha-dystroglycan (α-DG). α-DG is an important component of dystrophin–dystroglycan complex which links extracellular matrices with actin cytoskeleton and thereby maintains signalling cascades essential for the development of tissues and organs. The mutation T192M in α-DG hampers proper glycosylation of α-DG thereby developing limb girdle muscular dystrophy. Prototype virus for Old World Arenaviruses (OWV), Lymphocytic Choriomeningitis virus (LCMV) also uses this α-DG as host cell receptor and invades the host cell causing a disease called Lymphocytic choriomeningitis, an infection to meninges. Thereby, interaction of α-DG and LCMV has become an interesting object of study to predict the mode of the disease onset. In our current work, we have used homology modelling, molecular docking and molecular dynamics (MD) with temperature variation. We have identified significant structural differences between wild type (WT) and mutant (MT) α-DG in terms of spatiotemporal orientations of amino acids. This change in the folding patterns of the WT and MT α-DG has brought forth a different interaction pattern of the WT and MT α-DG with GP1 protein from LCMV as reflected in our docking simulations. Further MD simulations with the complexes over tropical and temperate environment have revealed that MT-α-DG-LCMV GP1 complex is relatively more stable than the wild type counterpart. It has also been found that LCMV GP1 has interacted strongly with mutant α-DG. Our studies therefore has shed light on the structure and molecular interaction pattern of LCMV with MT α-DG and also indicate a possibility of T192M mutant in α-DG making the receptor to interact strongly with LCMV GP1. These insights also provide clues to develop possible therapeutic approaches.

Keywords Homology modelling · Molecular dynamics · Limb girdle muscular dystrophy · Viral infection-drug design

S. Bhattacharya · S. Bhattacharjee · P.K. Biswas · A. Das · R. Dasgupta · A. Bagchi (✉)
Department of Biochemistry and Biophysics, University of Kalyani,
Nadia 741235, West Bengal, India
e-mail: angshuman_bagchi@yahoo.com

© The Author(s) 2015
N.B. Muppalaneni and V.K. Gunjan (eds.), *Computational Intelligence in Medical Informatics*, Forensic and Medical Bioinformatics,
DOI 10.1007/978-981-287-260-9_2

11

Abbreviations

DG   Dystroglycan
MDDGC9   Muscular Dystrophy, Dystroglycanopathy, Type C9
OMIM   Online Mendelian Inheritance in Man
PDB   Protein Data Bank
RMSD   Root Mean Square Deviation
WT   Wildtype
MT   Mutated

Highlights

1. T192M mutation causes changes in alpha-dystroglycan structure.
2. Interaction between Lymphocytic Choriomeningitis Virus (LCMV) with its receptor alpha-dystroglycan changes due to the mutation.
3. Molecular dynamics run with temperature variation shows interaction of LCMV with mutated receptor is stronger than the wild type receptor.

1 Introduction

Dystroglycan (DG) is a cell surface receptor and it belongs to dystrophin–dystroglycan complex which links extracellular matrices with cellular actin cytoskeleton [1]. Cellular actin cytoskeleton plays crucial role in development of tissues, organs not only by providing mechanical strengths to cells but also by taking active part in regulating macromolecular events [2] required for proper development [3]. Matured DG protein is composed of two subunits: alpha-dystroglycan (α-DG), the extracellular part of DG that interacts and receives signals from extracellular proteins like laminin, parlecan, argin, etc. and beta dystroglycan (β-DG), which remains bound to cell membrane [1]. This β-DG is linked with cellular dystrophin, which is an active acin binding protein [4]. The protein α-DG contains a mucin like region sandwiched between its N terminal and C terminal globular parts. This mucin rich region is essential for the proper glycosylation and functioning of α-DG [1]. Reportedly, a point mutation T192M in α-DG causes limb girdle muscular dystrophy, MDDGC9 (OMIM: 613818) affecting brain and nervous system growth resulting in severe cognitive impairments, mental retardation and delayed motor development [5, 6]. This mutation actually hampers LARGE (a transacetylase) mediated glycosylation leading to the disease onset. This signifies the role of α-DG in development and muscle strength.

Interestingly, α-DG also serves as a receptor to a large class of arenaviridae (AV), Old World viruses (OWV) [7]. OWVs invade host system by its preglycoprotein complex which has three major components: N terminal stable signal peptide (SSP), followed by glycoprotein G1 (GP1 ∼ 42 kDa) and glycoprotein G2 (GP2 ∼ 35 kDa) domains [7]. These viruses interact with the N terminal few amino acids and mucin rich region of α-DG via their GP1 for attachment and fusion to the host system. Some of these AVs are also causative agents of diseases like lassa fever, Bolivian hemorrhagic fever, lymphocytic choriomeningitis (LCM), etc. causing significant mortality in affected human populations [8]. The prototype for OWV, Lymphocytic Choriomeningitis Virus (LCMV) causing LCM, a chronic asymptomatic, lifelong infection of meninges which is the membrane to protect central nervous systems [9, 10]. This has tempted us to study the molecular interaction pattern of α-DG and LCMV GP1. Although detailed studies have been carried out on LCMV regarding its global distribution, pathogenicity, residues essential for its interaction with α-DG [8], the insight to its molecular interaction with MT α-DG is still not deciphered. Therefore, we have ventured into the prediction of molecular aspects of disease propagation. We have taken a homology modelling approach to construct the model of LCMV GP1 domain and thereafter have docked this model with wild type (WT α-DG) and T192M mutant (MT α-DG) forms of α-DG separately. Our initial studies [6] with α-DG structures have also revealed that the mutant structure differs significantly in terms of its surface hydrophobicity in vicinity to the mutation site, 192. That is why we have focused on the structural changes occurring in the vicinity of mutation site and its effect on the binding orientation of LCMV GP1 with MT α-DG to elucidate the effect of the above-mentioned mutation. Molecular docking and subsequent molecular mechanistic simulations for the first time have revealed that MT α-DG has increased numbers of intermolecular hydrogen and hydrophobic bonds with LCMV GP1 establishing a stable interaction. Further we have employed MD with elevated temperature to monitor the changes in this interaction pattern. This study therefore for the first time sheds some light into the scenario where MT-α-DG interacts with LCMV. And we have observed the pattern indeed changed with mutated receptor protein. In future, this work will be helpful in designing some effective therapeutic approaches to defeat the virus borne disease.

2 Materials and Methods

2.1 Template Search

First, the sequence of LCMV GP1 domain, comprising amino acid residues 88–241, was extracted from UniprotKB (id: E2D674). This was then used as input sequence to search for the template in FUGUE [11] and pdb [12] using PSI-BLAST [13]. In both the cases, the A chain of crystal structure of envelope glycoprotein Machupo virus GP1 (PDB code: 2WFO) was obtained as the closest match.

2.2 Homology Modelling

This structure was then used as template to construct homology model for LCMV GP1 using modbase [14] web server with sequence-sequence, profile-sequence

(PSI-BLAST) and sequence-profile (Modeller Built-Profile) methods. Since the template is a model of New World virus- Machupo virus GP1, the modelling of our desired protein was not so straightforward. So, alternatively, model for this region (amino acid residues 88–241) has also been constructed using I-TASSER and RaptorX model building servers independently. Both these servers used 2WFO as the template. The models so generated were then superimposed on the previously built model to generate as consensus report (Supplementary File 1). All models were tested with Verify 3D [15] and PROCHECK [16]. We have also denatured the structure completely taking it up to 373 K starting from 273 K, then have allowed it to cool down to 273 K. And the structures were then superimposed in order to check for the folding accuracy. The Root Mean Square Deviation (RMSD) for these two superimposed files have been found to be 0.8 Å (Supplementary File 2) which implies that they are indeed identical and the folding patterns are conserved.

2.3 Molecular Docking

Next, we used this model to dock with WT α-DG and MT α-DG [6] separately using Z dock [17], PyDock [18] and Patchdock [19]. Docked complexes with best score and lower energy were selected and there were total 25 docked complexes (5 from Z dock, 10 from Pydock and 10 from Fire Dock) for each type of complex, i.e. WT-α-DG with LCMV and MT-α-DG with LCMV. Resultant docked complexes were then energy minimized keeping their backbones fixed to ensure proper interactions. 800 cycles of conjugate gradient energy minimization steps were applied with CHARMm [20] force field until the structure reached a final derivative of 0.01 kcal/mole. After each minimization, the structures were verified with Verify3D and PROCHECK. Further analyses of the docked complexes were done using Discovery Studio (DS) package version 2.5 and Protein Interaction Calculator [21]. Binding sites were detected with analyse binding site tool from DS for all the docked complexes to validate the docking process. Finally, the complexes with the lowest energy profile and with binding sites consensus with literature [8, 10] have been selected for wild type and mutated complexes. The complexes were then further subjected to minimization in explicit solvent system with CHARMm force field. Dielectric constant of the molecular surface was kept 80 as of periodic water bound condition.

2.4 Molecular Dynamics Simulations

Minimized complexes then were brought to elevated tropical temperatures (i.e. 300–320 K) and 10,000 time step of production run was performed in each of the cases keeping pressure constant. The conformational changes were recorded and plotted against time.

3 Results

3.1 The Conformation of MT α-DG Contributes to Altered Interaction with LCMV GP1

Our studies [6] with the models of both WT and MT α-DG initially have figured out that mutant structure has more buried surfaces than that of its wild type form. In order to investigate further the art of interactions, we have docked LCMV GP1 with WT and MT α-DG, separately. Analyses of docked pattern of the models clearly have reflected that there are indeed significant differences both in terms of amino acid residues involved in interaction as well as the spatiotemporal orientation of LCMV GP1, docked with either WT or MT α-DG (Fig. 1a). As mentioned earlier, the focus of our study was the immediate vicinity of mutation site at the amino acid residue position 192. And we have marked those amino acid residues 88–96 of LCMV GP1 have successfully interacted with a groove formed by amino acid residues 142–192 of α-DG (Fig. 1b). We have attempted to analyse the total number of intermolecular hydrogen bonds (H-bond) and hydrophobic interactions (HPI) formed between LCMV GP1 and α-DG in both the cases, in two different systems, i.e. discovery studio 2.5 and protein interaction calculator. Normalized average values of the bonds formed have been used to generate a graphical view (Fig. 1c) which reflected that LCMV-GP1 docked with MT α-DG had more numbers of H-Bonds and HPI interactions than the bonds formed for its docked model with WT-α-DG (left panel, Fig. 1c). Hydrogen bonds and hydrophobic interactions are the non-covalent interactions which are meant to stabilize the overall interaction occurring between two proteins [22]. Disturbance in these bonds will definitely affect the binding. Studies with the structures in further details have marked this dispute to be largely contributed by a huge structural change in mutant form of α-DG in vicinity to the mutation site (Met 192) that has resulted in a new space for LCMV GP1 interaction and also has created potentially altered kinks among few residues (arrows, Fig. 1d, e). Superimpositions of models of WT and MT-α-DG, minimized with either vacuum [6] or explicit solvent condition (Supplementary File 3) have revealed that there indeed exist structural differences between the two structures.

3.2 Identical Amino Acids Bent Differentially to Give a New Conformation to Mutant α-DG

Further assessment of the afore mentioned portion of α-DG has identified amino acid stretch 166–171 of α-DG to be the key to these disputes. The relative bends and positioning of these residues have formed a surface which is more closed or buried in MT α-DG than that of its wild type form (Fig. 2a). Surprisingly, dihedral angles between the adjacent residues in this patch, i.e. Ser166-Val167, Val167-Arg168,

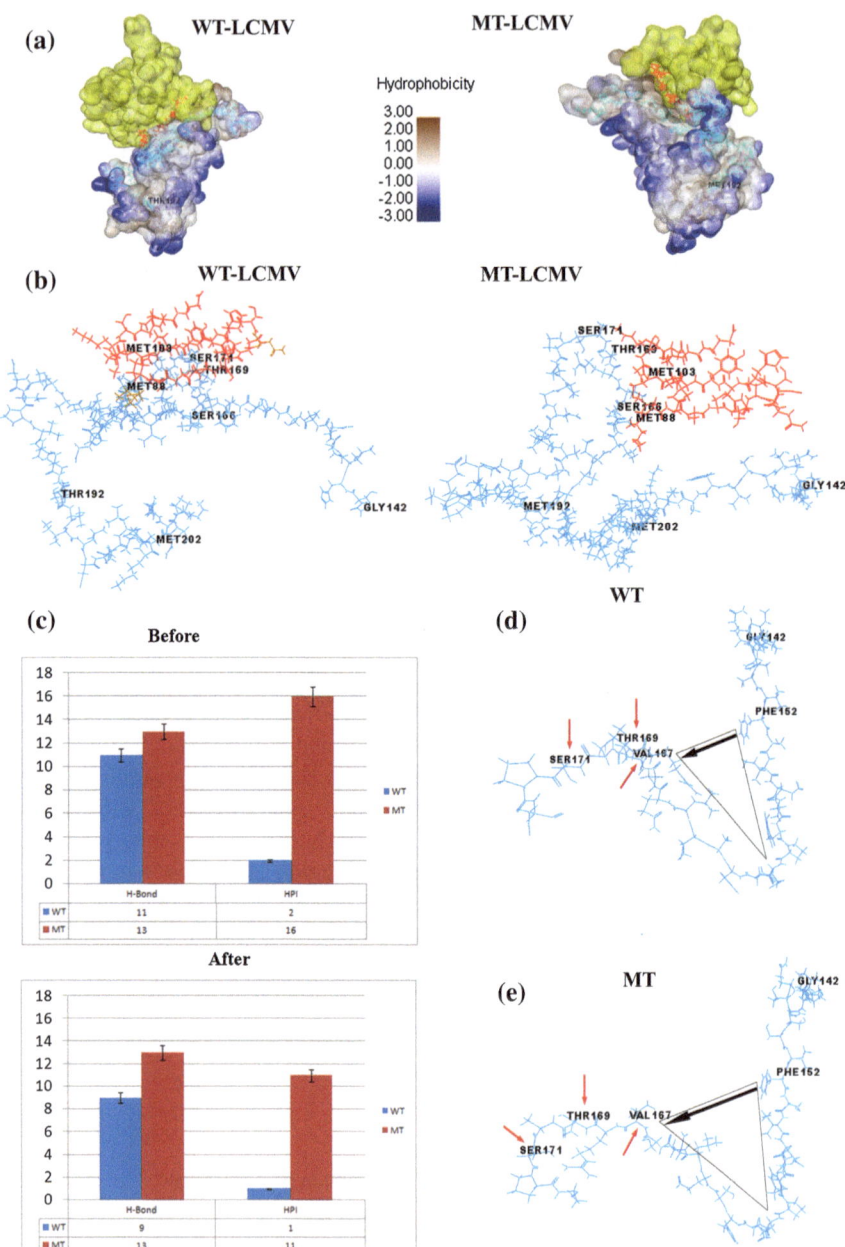

Fig. 1 LCMV GP1 exhibited a different spatiotemporal orientation when interacted with MT α-DG. **a** Docked structures of LCMV GP1 (*red backbone, yellow surface*) with WT-α-DG (*blue backbone, left panel*) and with MT α-DG (*blue backbone, right panel*). The common interacting potion has been shown with *red backbone* representation. **b** Amino acids in the vicinity of mutation site 192nd position and the different orientation of the docked complex. **c** Differences in intermolecular hydrogen bond and hydrophobic interactions have been shown. **d, e** Structural differences between WT and MT α-DG that open up an alternative binding site in case of mutated structure

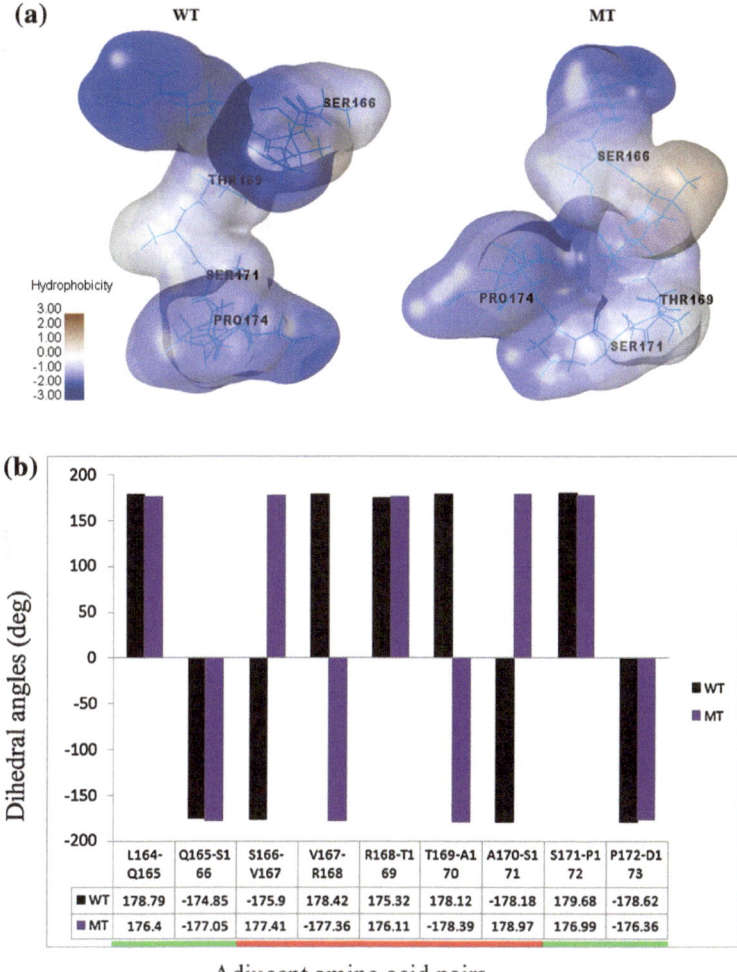

Fig. 2 Unusual kinks in amino acid pairs burying the surface in MT α-DG. **a** Different surface hydrophobicity between WT and MT α-DG making the interaction groove more buried. **b** The change in dihedral angles contributing to altered kinks in MT α-DG

Thr169-Ala170 and Ala170-Ser171, have possessed almost 180° change in two structures of α-DG (Fig. 2b). The corresponding changes of the angles are: Ser166-Val167 (−175.90) → (177.41), Val167-Arg168 (178.42) → (−177.36), Thr169-Ala170 (178.12) → (−178.39) and Ala170-Ser171 (−178.18) → (178.97) from WT → MT α-DG. Only amino acid pair Arg168-Thr169 (175.32 → 176.11) has similar dihedral angle in both the forms of α-DG.

3.3 Tropical Temperature Alters MT α-DG and LCMV GP1 Interactions

Lymphocytic choriomeningitis virus appears everywhere due to the large distribution of its host house mouse, rodents except for the Antarctica [23]. The serological prevalence of the virus has been reported to be centralized in Africa, but disease distribution occurs worldwide. Studies with patients suffering from lymphocytic choriomeningitis have revealed that this disease is prevalent in temperate zone [23]. Moreover, majority of Old World viral diseases occur at tropical and temperate regions [24]. So, we have employed MD run with a temperature range 27–47 °C so that we can monitor the changes in the interaction occurring between the receptor, α-DG and its ligand, LCMV-GP1. Changes in RMSD of the conformations over time in three different temperatures, i.e. 27, 37 and 47 °C (300, 310 and 320 K, respectively) have clearly demonstrated that the complex of MT-α-DG and LCMV-GP1 shows less alteration in RMSD with temperature change than the complex of WT-α-DG and LCMV-GP1 does. In other words, the interaction between MT-α-DG and LCMV-GP1 is much more stable than its wild type counterpart (Fig. 3).

4 Discussions

LCM, an infection of meninges which serve as a protective membrane for central nervous system, is developed due to the invasion of its causative agent LCMV to host system via α-DG, a host cell surface receptor and this interaction is largely counted on LARGE mediated proper glycosylation of matured α-DG. Additionally,

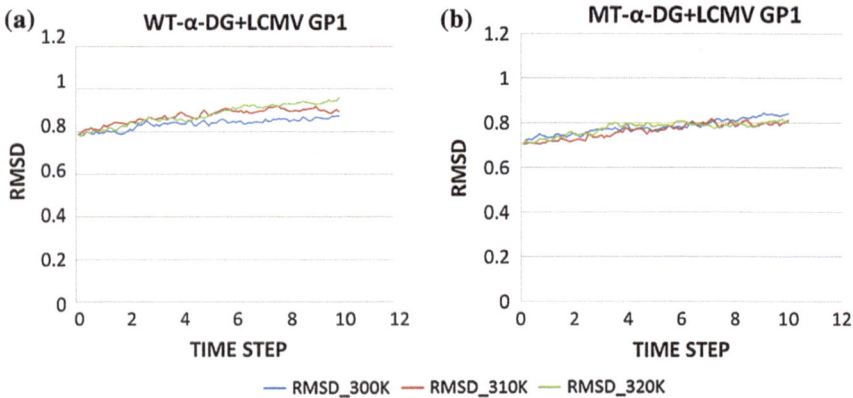

Fig. 3 Effect of temperature on the interactions. **a, b** RMSD versus Time plots exhibit that in case of the complex of LCMV GP1 with wild-type receptor has shown greater deviation in RMSD for the conformations (**a**). On the contrary, complex of LCMV GP1 with the mutated receptor has less conformational changes (**b**) with change in temperatures

a mutant form of this α-DG (T192M) hampers LARGE mediated glycosylation resulting in altercation in signalling cascade essential for proper brain development. Patients with T192M mutation in α-DG develop severe mental retardations along with cognitive impairment. These common features of the two diseases, i.e. dependence on LARGE, involvement of α-DG, proper glycosylation and above all affecting nervous system, brain, have made the interaction of LCMV and α-DG an invincible study object to us. In our current work, we have found that amino acid stretches 166–171 of α-DG in the close vicinity to mutation site of Met 192 have contributed to structural bends which finally have created relatively more buried conformation for MT α-DG than that of wild type one. Majority of amino acid pairs have almost 180° orientation difference in their dihedral angles which imply that replacement of polar Thr 192 with non-polar Met has conferred structural strains to the positional orientations of its upstream and downstream amino acids. Because of this, mutant form of α-DG has possessed a buried conformation with those amino acid stretches which otherwise have a widened window in the wild type α-DG. And this has forced LCMV GP1 to occupy a different space for interaction with MT α-DG. But this has been proven to be advantageous for the interaction of the two participating proteins since interaction of LCMV GP1 with α-DG has more main backboned based H-Bonds formed, compared to its interaction with WT-α-DG. And these same results have also been reflected with solvent explicit models. Interestingly, this result has an indication of likelihood of susceptibility to LCMV infection in the presence of the T192M mutation. This has urged us to look for the possible fold pattern of LCMV GP1 that can serve as a potential blocking site to restrict the disease propagation. In our study, as we already have shown that stretch of amino acids 88–96 of LCMV GP1 has remained a common mediator of interactions for both WT and MT α-DG and it also has possessed a particular fold pattern. In future, analysis of this structure will enlighten a path for developing novel therapeutic approaches against this viral transmission. Moreover, temperature-based variation on the conformations from our MD run have revealed that the interaction of LCMV GP1 and α-DG with T192M mutation is significantly stable than the interaction of LCMV GP1 with the wild type receptor. In conclusion, our experiments have shed lights on the molecular interactions of LCMV GP1 with MT α-DG, for the first time of its kind. Experimental results, here, have shown that structural aberrations in mutant α-DG owing to the altered bends in its amino acids residues have lead to changes in its interaction pattern with LCMV GP1. But this changed spatiotemporal orientation finally has resulted in strong binding of LCMV GP1 with MT α-DG, compared to WT-α-DG as is revealed by MD simulations and hydrogen bond analysis. This indicates a likelihood of LCMV susceptibility when α-DG possesses T192M mutation. T192M mutation is a naturally occurring mutation in human causing structural changes of the protein and subsequently diminishing the effective signal transduction. Blocking the activity of this mutant form with small antagonists may anyway hamper the process. Gene therapies that deliver functional transcript of the protein may ameliorate the ill effect of the mutation. But from this study, we have found that LCMV GP1 has stronger interactions with the mutated form than the wild type one. That implies an effective therapeutic strategy should be developed against the viral protein to block the interaction of LCMV GP1 with the host receptor protein. Studying the particular folding pattern (s) of LCMV GP1 in near future may lead the path to develop new potential therapeutic approaches to bar the spread of this viral transmission. The only licensed drug available to certain diseases caused by Arenaviruses is nucleoside analogue ribavirin [25]. A detailed analysis of this interaction with total protein model will generate deeper insight to the molecular basis of the viral attachment to mutant protein. Our future study therefore includes (i) homology modelling of total α-DG with its mutant form as well as that of LCMV GP1, (ii) docking simulations followed by thereafter analyses of the interactions and (iii) small molecular library screening to find effective molecule against the viral GP1.

Acknowledgments Authors are thankful to Dept of Biochemistry and Biophysics, University of Kalyani for their continuous support and for providing the necessary instruments to carry out the experiments. The authors would like to thank the ongoing DST-PURSE programme. SB and AD also are thankful to UGC, India and CSIR, India for their respective fellowships, and the DBT (project no. BT/PR6869/BID/7/417/2012) for the necessary infrastructural support.

Conflict of Interest
The authors declare no conflict of interest.

Appendix

Supplementary
File 1 Superimposed models built from Modbase, RaptorX and I-TASSER

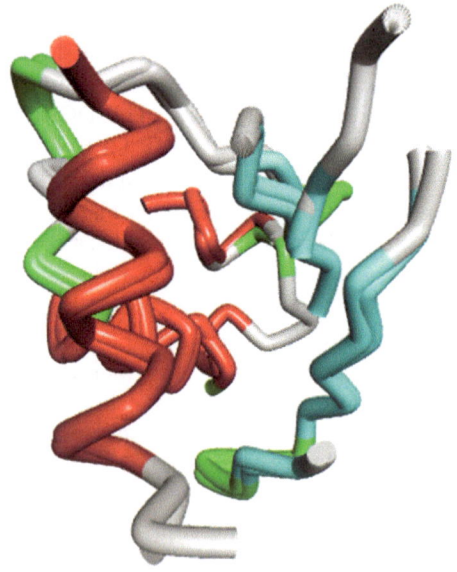

**Supplementary
File 2** Superimposed
Reformed structure with its
original structure after heat
denaturation

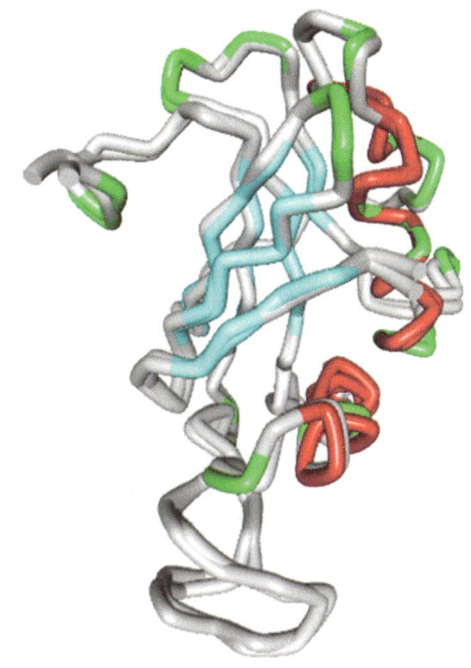

**Supplementary
File 3** Superimposed WT
and MT alpha-dystroglycan
minimized with explicit
solvent system

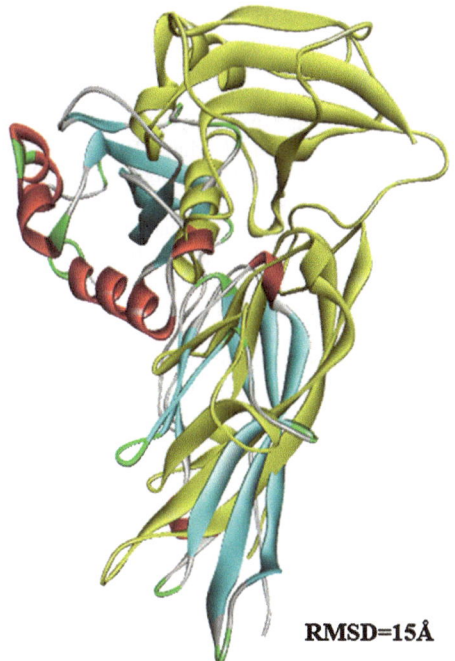

RMSD=15Å

References

1. Henry MD, Campbell KP (1999) Dystroglycan inside and out. Curr Opin Cell Biol 11:602–607. http://dx.doi.org/10.1016/S0955-0674(99)00024-1
2. Pollard TD (1986) Mechanism of actin filament self-assembly and regulation of the process by actin-binding proteins. Biophys J 49:149–151. http://dx.doi.org/10.1016/S0006-3495(86) 83630-X
3. Cheever TR, Ervasti JM (2013) Actin isoforms in neuronal development and function. Int Rev Cell Mol Biol 301:157–213. http://dx.doi.org/10.1016/B978-0-12-407704-1.00004-X
4. Henry MD, Campbell KP (1996) Dystroglycan: an extracellular matrix receptor linked to the cytoskeleton. Curr Opin Cell Biol 8:625–631. http://dx.doi.org/10.1016/S0955-0674(96) 80103-7
5. Dinçer P, Balci B, Yuva Y, Talim B et al (2003) A novel form of recessive limb girdle muscular dystrophy with mental retardation and abnormal expression of alpha-dystroglycan. Neuromuscul Disord 13:771–778. http://dx.doi.org/10.1016/S0960-8966(03)00161-5
6. Bhattacharya S, Das A, Ghosh S, Dasgupta R, Bagchi A (2014) Hypoglycosylation of dystroglycan due to T192M mutation: a molecular insight behind the fact. Gene 537:108–114. http://dx.doi.org/10.1016/j.gene.2013.11.071
7. Spiropoulou CF, Kunz S, Rollin PE et al (2002) New world arenavirus clade C, but not clade A and B viruses, utilizes alpha-dystroglycan as its major receptor. J Virol 76:5140–5146
8. Oldstone MB, Campbell KP (2011) Decoding arenavirus pathogenesis: essential roles for alpha-dystroglycan-virus interactions and the immune response. Virology 411:170–179. http://dx.doi.org/10.1016/j.virol.2010.11.023
9. Lapošová K, Pastoreková S, Tomášková J, Lymphocytic choriomeningitis virus: invisible but not innocent. Acta Virol 57:160–170. http://dx.doi.org/10.4149/av_2013_02_160
10. Kunz S, Sevilla N, McGavern DB et al (2001) Molecular analysis of the interaction of LCMV with its cellular receptor [alpha]-dystroglycan. J Cell Biol 155:301–310. http://dx.doi.org/10.1083/jcb.200104103
11. Shi J, Blundell TL, Mizuguchi K (2001) FUGUE: sequence-structure homology recognition using environment-specific substitution tables and structure- dependent gap penalties. J Mol Biol 310:243–257. http://dx.doi.org/10.1006/jmbi.2001.4762
12. Berman HM (2008) The protein data bank: a historical perspective. Acta Crystallogr A 64:88–95. http://dx.doi.org/10.1107/S0108767307035623
13. Altschul SF, Gish W, Miller W et al (1990) Basic local alignment search tool. J Mol Biol 215:403–410. http://dx.doi.org/10.1006/jmbi.1990.9999
14. Pieper U, Webb BM, Barkan DT et al (2011) ModBase, a database of annotated comparative protein structure models, and associated resources, Nucleic Acids Res 39(Database issue): D465–D474. http://dx.doi.org/10.1093/nar/gkq1091
15. Eisenberg D, Lüthy R, Bowie JU (1997) VERIFY3D: assessment of protein models with three-dimensional profiles. Methods Enzymol 277:396–404. http://dx.doi.org/10.1016/S0076-6879(97)77022-8
16. Laskowski RA, MacArthur MW, Moss DS et al (1993) PROCHECK—a program to check the stereochemical quality of protein structures. Appl Crystallogr 26:283–291. http://dx.doi.org/10.1107/S0021889892009944
17. Chen R, Li L, Weng Z, ZDOCK: an initial-stage protein-docking algorithm. Proteins 52:80–87. http://dx.doi.org/10.1002/prot.10389
18. Jiménez-García B, Pons C, Fernández-Recio J (2013) pyDockWEB: a web server for rigid-body protein-protein docking using electrostatics and desolvation scoring. Bioinformatics 29:1698–1699. http://dx.doi.org/10.1093/bioinformatics/btt262
19. Schneidman-Duhovny D, Inbar Y, Nussinov R, Wolfson HJ (2005) PatchDock and SymmDock: servers for rigid and symmetric docking. Nucl Acids Res 33:W363–W367

20. Brooks BR, Bruccoleri RE, Olafson BD et al (1983) CHARMM: A program for macromolecular energy, minimization, and dynamics calculations. J Comput Chem 4:87–217. http://dx.doi.org/10.1002/jcc.540040211

21. Tina KG, Bhadra R, Srinivasan N (2007) PIC: Protein interactions calculator. Nucleic Acids Res 35(Web Server issue):W473–W476. http://dx.doi.org/10.1093/nar/gkm423

22. Jaenicke R (1991) Protein stability and molecular adaptation to extreme conditions. Eur J Biochem 202:715–728

23. The Centre for Food Security and Public Health (2010) Institute for International Co-operation and in Animal Biologics. Iowa State University, Ames. http://www.cfsph.iastate.edu/Factsheets/pdfs/lymphocytic_choriomeningitis.pdf

24. Pasqual G, Rojek JM, Masin M, Chatton JY, Kunz S (2011) Old world arenaviruses enter the host cell via the multivesicular body and depend on the endosomal sorting complex required for transport. PLoS Pathog 7:e1002232. doi:10.1371/journal.ppat.1002232

25. Lee AM, Pasquato A, Kunz S (2011) Novel approaches in anti-arenaviral drug development. Virology 411:163–169

Structural and Functional Characterization of *Arabidopsis thaliana* WW Domain Containing Protein F4JC80

Amit Das, Simanti Bhattacharya, Angshuman Bagchi and Rakhi Dasgupta

Abstract WW domains are the smallest known independently foldable protein structural motifs that are involved in cellular events like protein turnover, splicing, development, and tumor growth control. These motifs bind the polyproline rich ligands. While the WW domains of animal origin are well characterized, the same from plant origin are not well documented yet. Despite the small repertoire of WW proteome of plants (in comparison to animal WW proteome) functional diversity is reported to be equally vivid for plants also. Here, for the first time, we report the structural and functional properties of an *Arabidopsis thaliana* (*At*) WW domain containing protein F4JC80 by using homology modeling and docking techniques. Our findings report that the *At* F4JC80 protein contains two WW domains which bear the standard triple β sheet structure and structurally and functionally resemble Class I WW domains of E3 ubiquitin ligase family but their structural differences impact their polypeptide binding abilities differently.

Keywords WW domain · *Arabidopsis thaliana* · Transcription · Splicing · Homology modeling · Molecular docking

1 Introduction

Different cellular phenomena like protein ubiquitination, splicing, organ development, tumor progression, and suppression are associated with large multi-protein interactions. Conserved patches of proteins, known as domains, often assign the structural and functional properties of a protein. Due to their functional importance,

A. Das · S. Bhattacharya · A. Bagchi (✉) · R. Dasgupta (✉)
Department of Biochemistry and Biophysics, University of Kalyani,
Kalyani, Nadia 741235, WB, India
e-mail: angshuman_bagchi@yahoo.com

R. Dasgupta
e-mail: rdgadg@gmail.com

N.B. Muppalaneni and V.K. Gunjan (eds.), *Computational Intelligence in Medical Informatics*, Forensic and Medical Bioinformatics,
DOI 10.1007/978-981-287-260-9_3

evolutionary forces often enable protein domains to accommodate more sequence variations keeping the overall structural folding pattern intact. One such great example is the WW domain which is known to be the smallest protein module (only 30–40 amino acids long) that is able to fold independently in solution [1]. This domain harbors its name from the presence of two conserved tryptophan residues residing toward the two termini of this 30 amino acids stretch [2]. Other than the two conserved residues, there are 2–3 aromatic amino acids, mostly tyrosine, are found in the middle of this domain and a conserved proline residue is located at the end of this domain [1]. Other than these 4–5 conserved residues, the remaining parts of these WW domains are found to be highly variable in nature. Despite this huge sequence variation, this small 30–40 amino acids stretch of a WW domain is always found to fold into a triple β sheet structure where the individual β sheets are generally connected by small β turns [1, 2]. However, the lengths of the individual β sheets vary greatly among different WW domains. Generally, proteins or peptides containing polyproline rich regions act as ligands of this small protein module. Different WW domains interact with polyproline rich ligands or polypeptides (PRL/PRP) of different compositions and lengths. Based on their PRL/PRP choices, WW domains are classified into five different classes [1]. Among these five classes, class I WW domain comprises of E3 ubiquitin ligases like NEDD4, ITCH as well as transcription regulator like YAP1 and oxidoreductase WWOX (NEDD4: E3 ubiquitin-protein ligase NEDD4/Neural precursor cell expressed developmentally down-regulated protein 4; ITCH: E3 ubiquitin-protein ligase Itchy; YAP1: Yorkie homolog/65 kDa Yes-associated protein; WWOX: WW domain containing oxidoreductase; APBB1: Amyloid beta A4 precursor protein-binding family B member 1; FBP/FNBP: Formin binding protein; WBP: WW domain binding protein). Proteins like FBP11 and Prp40 which are part of the eukaryotic splicing machinery form the Class II and Class V, respectively, while prolyl isomerases like Pin1 is the sole member of the Class IV. Other WW domain containing proteins like FE65 (also known as APBB1) and FBP21 (also known as WBP4) form the class III WW domains. Salah et al. [1] in their review about WW domains have mentioned in details about the specific ligand choices of each class of WW domains [3–8].

Irrespective of this small length, the functional impact of WW domain is enormous and it is found to be associated with all the above-mentioned cellular phenomena. It is often associated with large cellular protein networks. Diseases like Alzheimer and Huntington disease are also associated with different WW domain containing proteins (WW-CP) who's altered or impaired interactions are associated with these disease prognosis [1]. Like the functional diversity the distribution and architectural diversity of WW domains are also significantly diversified. The pfam WW domain family (http://www.pfam.xfam.org) consists of 8,927 WW domain sequences form 357 species with 403 different architecture.

Over the last decade, the WW domain has been structurally and functionally well characterized. Unfortunately most of these studies whether cellular, biochemical, structural, or at organism level, have been associated with WW domains of either animals or fungi. Studies about plant WW domain is restricted to a handful of articles despite the findings that WW domains of plants are also associated with diverse cellular events like inhibition of plant virus replication, cuticle development, plant flowering time control, morphological development, RNA processing, and prolyl isomerization [9–14]. While the last two functions are quite common to the WW domain family, others are specific to plants. Considering the conserveness of different pathways involving RNA processing and prolyl isomerization, it would not be surprising to find out the involvement of WW domains in similar events of plants. In this context, it is also surprising that the functional diversity of the plant WW domains have not been well addressed yet. While plants are much less diversified in terms of their WW proteome (Table 1) when compared to that of animals, the WW domain-related studies of model plants like *Arabidopsis thaliana* (*A. thaliana*) and *Oryza sativa* (*O. sativa*) are significantly limited. Without better understanding of the plant WW domain functions and their involvement in different cellular events, it would not be possible to specifically categorize the similarities and differences of plant and animal WW domains, their PRL/PRP preferences as well as their respective protein interactions. Especially characterization of those WW domains and WW-CPs that are involved in evolutionary conserved processes like RNA processing, prolyl isomerization, protein turnover would provide significant insight into better understanding of these essential cellular processes that encompass all kinds of eukaryotes. In this context, we have undertaken homology modeling- and docking-based studies to characterize and functionally annotate the uncharacterized WW-CPs of model plants like *A. thaliana*.

As described in Table 1, the *A. thaliana* WW proteome consists of only 28 proteins (22 unique sequences and six redundant sequences) which has a total of 41 WW domain sequences. Among these 22 unique sequences, 8 are yet to be functionally annotated while others were reported to be involved in most of the above-mentioned plant WW domain functions. Among these eight uncharacterized proteins, this study reports about the structural and functional characterization of

Table 1 Comparison of WW-CP proteome of plants and animals

Serial no.	Species name	Number of proteins with WW domains	Number of total WW domains
1	*Homo sapiens*	204	314
2	*Mus musculus*	139	243
3	*Danio rerio*	108	174
4	*Drosophila melanogaster*	76	139
5	*Arabidopsis thaliana*	28	41
6	*Oryza sativa*	16	25

According to the pfam database, WW domain containing proteins are more in number in animals than to the same of plants. While vertebrates (like human, mouse, or zebra fish) contain more than 100 WW domain containing proteins and even larger amounts of WW domain sequences (on average 1.5 WW domains per protein), variation of plant (*A. thaliana and O. sativa*) WW domains is much less, even in comparison to Arthropods (fruit fly). However, the functional diversity of plants and animals are comparable (details in text)

one such *A. thaliana* WW-CP, annotated as 'WW domain containing protein' (Uniprot ID: F4JC80) which harbors two WW domains. (During the rest of the article, this protein will be mentioned as *At* F4JC80).

2 Materials and Methods

The amino acid sequence of *At* F4JC80 (Uniprot KB ID: F4JC80, NCBI Accession number: NP_001154613.1) was retrieved from Uniprot KB (http://www.uniprot.org) and its corresponding conserved domain regions were identified via pfam [15]. The pfam output showed the presence of two WW domains in *At* F4JC80. The pfam identified WW domain regions were further used for homology modeling and subsequent docking and functional annotation studies.

For structural and functional characterization of a protein, it is essential to generate stereo-chemically fit structures that can be further used in different studies. Although there are couple of models of WW domains available at the Swiss Model repository for the protein of our interest, none of the models were found to be stereochemically fit. Thus, fresh models for the WW domains of *At* F4JC80 were required to be built. To identify potential templates for homology modeling of the *At* F4JC80 WW domains, NCBI PSI-BLAST (http://blast.ncbi.nlm.nih.gov/Blast.cgi?PAGE= Proteins) was performed against the protein databank (PDB) (http://www.rcsb.org) using the *At* F4JC80 WW domain region sequences as query sequences. For the BLAST search the default parameters were used (BLOSUM62 matrix, Word size: 3, Expect threshold: 100, Gap costs: Existence: 11, Extension: 1). However, low-sequence similarity and poor e-values of the PDB PSI-BLAST outputs rendered no direct choice of structures that can be used for homology modeling. The small length and conserved folding patterns of WW domains often enables us to produce stereochemically fit structure even with templates of low similarity. So we have used the modbase [16] server (slow mode) for homology modeling of *At* F4JC80 WW domains. Modbase while generating the homology models identifies the best template for homology modeling. Modbase generates models using ModPipe, which is an automated modeling pipeline. ModPipe uses Modeler for fold assignment, sequence–structure alignment, model building, and model assessment. The validations of the structures obtained as outputs from Modbase were performed using the SAVES (v3) server with Procheck, Verify_3D, and Errat analysis [17–19] as well as through the ProSA [20] tool. Model visualization, inspection, analysis, and figure preparation were performed using Discovery Studio Client 3.5 [Accelrys, United States, presently BIOVIA, Dassault systems, France]. Built models were further subjected to energy minimizations. The energy minimization processes were run with conjugate gradient algorithm for 4,000 steps by applying the CHARMm [22] force field until the structures reached the final derivative of 0.01 kcal/mole using the Discovery Studio 2.5 package [Biovia].

Functional annotations of the *At* F4JC80 WW domains were performed by structurally comparing the models with structures of known function through the

DALI server [21]. Only the outputs with a RMSD of <1.5 Å were used to assign functions to our domains of interest.

The central groove created by the three β sheets form the PRL/PRP binding region of any known WW domain [1] and till date all the known PRLs/PRPs of WW domains have been reported to be binding to this groove. In case of our docking studies, only those docked complexes where the PRL/PRP was found to be docked into this groove were considered for further analysis. Any other docking poses of ligand-receptor complex were rejected.

We have extracted the 17 different PRL/PRP structures from different known complexes of proteins with WW domains as receptors bound to different ligand molecules or polypeptides which are available at the Protein Data Bank (http://www.rcsb.org). These ligand structures were further heated to 373 K and then cooled to 273 K to release the structural constraints, if any, that might be present in them due to crystallization conditions or other structural limitations from the reported structural studies. These heating and cooling procedures were performed using the Accelrys Discovery Studio 2.5 package (Accelrys, USA; presently BIOVIA, Dassault Systems, France). Required energy minimization for these ligand structures were performed in Discovery Studio 2.5 using the Conjugate Gradient algorithm. All the docking studies were performed through the Z-DOCK server [23] (for rigid docking). Only those docked complexes that were found to have least energy as well as stereo-chemically fit, were selected for further refinement. These reliable best outputs of the Z-DOCK server were subjected to further refinement via the FlexPepDock server [24]. Since 10 different output models are generated by FlexPepDock, similar criteria as of selecting the Z-DOCK output were also applied to choose the output of the FlexPepDock server. Validations of the docked complexes were performed through SAVES server using previously mentioned analytical tools. Energy minimizations of the docked complexes were performed as mentioned above with backbone fixation to ensure proper interactions of the docked complexes. After each minimization, the structures of the complexes were checked through Verify_3D and Procheck to determine their stereo-chemical fitness.

3 Results and Discussions

3.1 Sequence Analysis

The sequence of *At* F4JC80 is 892 amino acids long. It harbors two WW domains (Fig. 1a). The N terminal WW domain is located between amino acids 186 and 213 whereas the other WW domain is located at extreme C termini of the protein (860–890 amino acids) (Fig. 1a). Both of these pfam identified WW domains were found to contain the two conserved Trp residues as well as two aromatic Tyr residues in the middle of each region and the conserved terminal proline residue

Fig. 1 Domain organization of *At* F4JC80 WW domains and structural properties of these WW domains (*left panel* for 1st WW domain and *right panel* for 2nd WW domain). **a** 892 amino acids long *At* F4JC80 harbors two WW domains (marked with *green* boxes with their amino acid positions marked on each side). The pfam match qualities of these two WW domains are shown in details. **b** The triple β sheet structure of the generated models of these two WW domains with each sheet numbering starting from N terminal. **c** Side view of the models shows the difference in the *upper* concave ligand binding groove. **d** ProSA plot of the energy functions of these two WW domains validates their structural properties. **e** and **f** Structural alignment of these two models with their templates shows the areas of differences (template shown in *yellow*, major differences are marked with *red arrow* head). **g** Sequence view of these model and template structures alignment

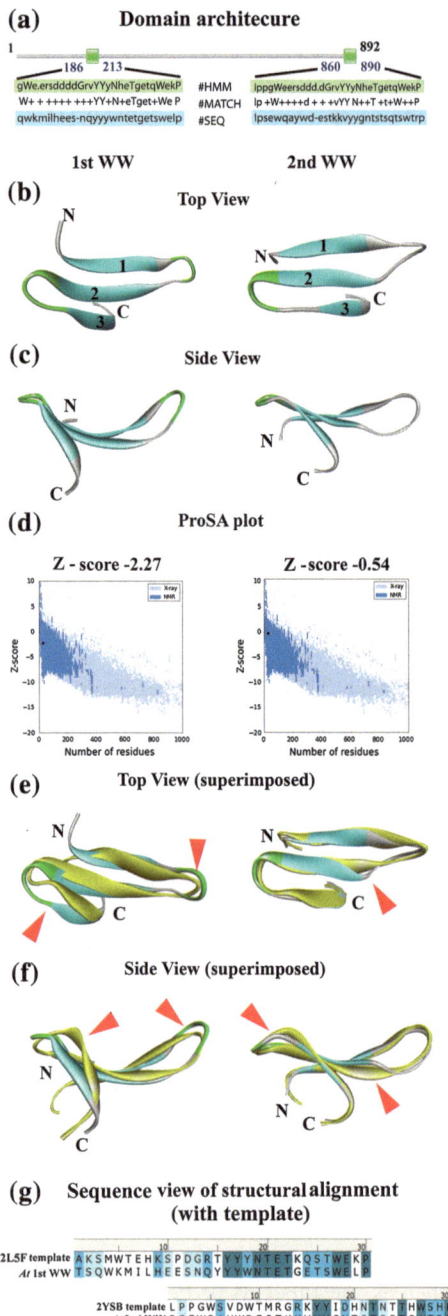

(Fig. 1a). The level of confidence to identify these two regions as WW domains were very high ($2.4e^{-07}$ for the 1st WW domain and $8.9e^{-06}$ for the 2nd WW domain of *At* F4JC80). Since protein domains are the most important region of any protein in terms of mediating its interaction with ligands, substrates, or other proteins; presence of only WW domain in *At* F4JC80 rendered these two WW domains as the sole functional motif of *At* F4JC80. So for further understanding of the functional properties of *At* F4JC80, these two WW domains were subjected to homology modeling.

Another important feature of *At* F4JC80 is the presence of polyproline rich stretches in this protein. PRLs are known as the ligands for WW domains and based on the composition of PRLs/PRPs, WW domains are classified into five classes [1]. The *At* F4JC80 contains four polyproline stretches among which three are four proline residues long (537–540aa, 546–549aa, and 568–571aa) while one is five proline residues long (556–560aa). Presence of both ligand and receptor domain (i.e., polyproline stretches and WW domains, respectively) could have significant impact in determining the state of activation (through the intra-molecular interactions of these ligand and domain), oligomerization (through mediating inter-molecular interactions), or binding partners (through interaction with other WW domains or polyproline stretches of other proteins) of *At* F4JC80.

3.2 Homology Modeling: Modeled Structure Description

3.2.1 1st WW Domain

The 1st WW domain of *At* F4JC80 was build using the structure having **PDB ID 2L5F**—chain A as template. The stereo-chemically fit model consists of three β sheets which are connected by β turns (Fig. 1b, left panel). The length of each of the three β sheets is different from each other. The side view of the *At* F4JC80 1st WW domain model shows a deep concave groove (Fig. 1c, left panel) which represents the primary PRL/PRP binding region of any WW domain. Along with Ramachandran plot (via Procheck), Verify_3D, and Errat analysis (See Appendix for details of about model quality), ProSA analysis of the model also validated its structural and energy-related properties (Fig. 1d, left panel). Superimposition of the model with its template identified the regions of deviations (arrow heads in Fig. 1e, f, left panel) which resulted in a RMSD of 1.79Å between the model and its template. This difference in RMSD may indicate slight difference from its template but it is mainly due to low sequence similarity with the template (Fig. 1g, left panel). However, this structurally validated model showed 80 % structural overlap with its template as well an overall Z-score of −2.27 which confirmed the good overall quality of the model which was also represented as a point in the plot (Fig. 1d, left panel) that is in a range solved by X-ray and NMR studies.

3.2.2 2nd WW Domain

The 2nd WW domain of *At* F4JC80 (modeled with **PDB ID: 2YSB** chain A as template) is also made of three variable length β sheets with a β turn connecting sheet 2 and 3 (Fig. 1b, c, right panel). Unlike the model of the 1st WW domain, sheet 1 and 2 of the 2nd WW domain model is connected by a longer loop structure and compared to the 1st WW domain model, the PRL/PRP binding groove of the 2nd WW domain is shallower (Fig. 1c, right panel). However, this model was also found to be stereo-chemically fit and the ProSA energy function analysis of the model rendered an overall Z-score of −0.54 which also falls in a range solved by X-ray and NMR techniques. Superimposition of this model structure (Fig. 1e–g, right panel) with its template was found to be more accurate (96.67 % structural overlap to template) than the 1st WW domain model and this also reflected in the lower RMSD of 0.74 Å.

Both the models of the 1st and 2nd WW domain of *At* F4JC80 were found to be structurally suitable and they represented the overall common architecture of WW domains. Even with similar architectures, WW domains are quite different in terms of their functions and PRL/PRP choices. We have used the two above-mentioned models to perform functional annotations to each of them.

3.3 Ligand Binding Properties and Functional Properties of Arabidopsis thaliana F4JC80 WW Domains

WW domains bind to polyproline rich stretches and their classification is also based on their PRL/PRP preferences. To understand the functional properties of the *At* F4JC80 WW domains, we performed docking of the generated models with the different PRLs/PRPs that are available at PDB. Refinement of the docked structure removed the structural constrains of the ligands while increasing its docking parameters (Fig. 2). The deep ligand binding groove of the 1st WW domain model was found to negatively impact the docking of the *At* F4JC80 1st WW domain (Fig. 2a, b, Table 2) with different PRLs/PRPs. Especially it affected the formation of Hydrogen bonds (H-bonds) in the deepest region of the groove by imposing distance constrains and thus no H-bond was found in the middle region of the ligand that spans over the triple β sheet structure. However, the 2nd WW domain was found to be more general in terms of its PRL/PRP choices (Table 2). Especially the shallow ligand binding groove (Fig. 2c, d) also contributed significantly toward interacting with PRLs/PRPs in the form of forming more hydrogen bonds in this region. Making the PRL/PRP structure flexible via Flexpepdock [24], this server have put the PRL/PRP in proper pose where it can better interact with the receptor WW domain (Fig. 2e, f).

Docking studies showed that both the WW domains mainly bound to the PRLs/PRPs of class I WW domains (comprising of WW domains from E3 ubiquitin ligases like NEDD4, ITCH, and YAP as mentioned earlier) [1]. Beyond this, the 2nd WW domain also showed preference toward PRLs/PRPs that primarily interacts with Class V WW domains (which are formed by WW domains of transcription factor like Prp40) [1].

Fig. 2 Docking of *At* F4JC80 WW domains with their ligands. **a, b** 1st WW domain with ligand from PDB ID: 2JO9 structure, (**c–f**) 2nd WW domain with ligand from PDB ID: 2V8F structure. **a–d** Receptor WW domain and ligand regions is shown as *ribbon* and Hydrogen bonds are shown in *green dashed lines*. **c, d** The interacting surface (according to hydrophobicity) is shown for the 2nd WW domain. (Due to interference of the clear visibility of H-bonds, similar surface is not shown for the 1st WW domain). **e, f** Front and side view, (respectively), of the flexpepdock output of the 2nd WW and its ligand interaction which highlights the variation in the ligand backbone to maximize the docking qualities. (For simplicity of representation, only one such flexible docking output is shown but similar approaches to maximize docking qualities were taken for all combinations of successful Z-DOCK output with both *At* F4JC80 WW domains and their ligands)

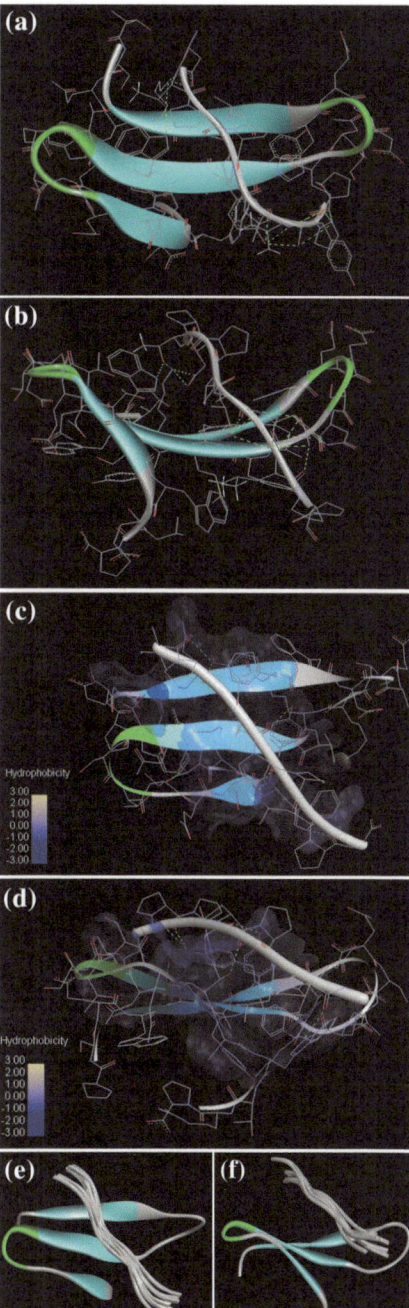

Table 2 Ligand binding choices of *At* F4JC80 WW domains

Serial no.	Ligand PDB ID	Ligand structure receptor protein	Ligand sequence	At 1st WW domain	At 2nd WW domain
1	1I5H	NEDD4-WW1	GSTILPIPGTPPPNYDSL	N/A	N
2	1K5R	YAP1	GTPPPYTVG	N/A	Y
3	2DJY	SMURF2-WW3	GPLGSELESPPPPYSRYPMD	N/A	Y
4	2DYF	FBP11-WW1	GSTAPPLPR	N/A	N
5	2EZ5	NEDD4-WW3	TGLPSYDEALH	N/A	Y
6	2HO2	FE65	PPPPPPPPL	N/A	Y
7	2JMF	Su(dx)-WW4	GPLGSPNTGAKQPPSYEDCIK	N/A	Y
8	2IO9	ITCH-WW3	EEPPPPYED	Y	Y
9	2JUP	FBP28-WW2	PPLIPPP	Y	Y
10	2KQ0	NEDD4-WW3	ILPTAPPEYMEA	Y	Y
11	2KXQ	SMURF2-WW2	GPLGSELESPPPPYSRYPMD	N/A	Y
12	2LAW	YAP1	TPPPAYLPPEDP	N/A	Y
13	2LB1	SMURF2-WW2	DTPPPAYLPPEDP	Y	Y
14	2LB2	NEDD4L-WW2	ETPPPGYLSEDG	N/A	Y
15	2RLY	FBP28-WW2	PTPPPLPP	N/A	Y
16	2RM0	FBP28-WW2	PPPLIPPP	N/A	Y
17	2V8F	PROFILIN	IPPPPLPGV	N/A	Y

Although WW domains form a triple β sheet structure, connected by β turns, variation of sequences result in their variable ligand choices. We have used online docking server Z-DOCK to find out the ligand binding ability of both the WW domains of *At* F4JC80 with 17 different ligand sequences extracted from already reported structures of different WW domain—ligand complexes (taken from protein database: http://www.rcsb.org). The result clearly shows that the 1st WW domain of *At* F4JC80 can bind to only a handful of ligands while the 2nd WW domain binds to almost all the types of proline rich ligands that were taken into consideration in this study. Since WW domains bind to their ligands only through their well known ligand binding groove (detailed in text), the deep groove that we have found in case of the 1st WW domain of F4JC80 could well be the reason for the inability of this domain to bind to different ligands as such deep grooves imposes a distance constrain to form successful atomic interactions

For better insight into the functional properties of the *At* F4JC80 WW domains, we have used the DALI server for functional assessment by structural comparison. The DALI output (only those outputs with a RMSD of <1.5 Å were considered) showed that both the *At* F4JC80 WW domains are structurally similar to the WW domains of NEDD4 family which also synergized with our findings from the docking studies.

Overall this study on the in silico characterization of *At* F4JC80 has identified the presence of two WW domains in this protein as well as assigned functional annotations two these domains by molecular docking techniques and structure-based functional assessment studies. Such computational studies which is a first of its kind on plant WW domains would provide a future direction on further characterization of this or similar types of proteins. Future studies addressing these proteins from experimental approach will prove to be significant in understanding the overall interactome of these proteins.

4 Conclusions

This study on structural and functional properties of *At* F4JC80 identified presence of two WW domains and four polyproline rich stretches in the protein. Both the models of the WW domain were found to contain the standard triple β sheet structure but there were significant difference in the PRL/PRP binding groove of the WW domains. This difference caused better ability of the 2nd WW domain to bind to PRL/PRPs as the deep groove was found to have a negative impact on the 1st WW domain's PRL/PRP binding capabilities. However, docking and structural comparison-based function annotation studies found both the WW domains belong to the Class I WW domains (specifically related to NEDD4 and SMURF family). Based on the well-characterized roles of NEDD4 and SMURF E3 ubiquitin ligases [4], it can be hypothesized that the *At* F4JC80 WW domains are most probably involved with gene expression control by regulating protein turnover during TGF β signaling. Also the presence of nuclear localization signal (regions 746-KRTKKK-751 795-KRKR-798 826-WREKVKRKRERAEKSQKKDPE–846, identified through NLStradamus server [25]), makes this protein an important player in nuclear context. Moreover, the ability of the 2nd WW domain to bind to ligands of WW domains belonging to Prp40 (PDB ID: **2DYF**) and ca150 (PDB ID: **2RM0**) as well as with formin polyproline region (PDB ID: **2VDF**) shows a wider arena of functional interactions of *At* F4JC80 in cellular context where it has the possibility to directly regulate cellular actin cytoskeleton [26] and gene transcription.

Acknowledgments The authors would like to thank the members of RD and AB laboratory for their continuous support and critical assessments. AD and SB would like to thank CSIR (India) and UGC (India), respectively, for their Ph.D. fellowships.

Author Contributions AD and SB have performed the modeling and docking studies and drafted the manuscript. RD and AB have analyzed the results and prepared the final version of the manuscript. All the authors have read and approved the final version of the manuscript.

Appendix

Details of *At* F4JC80 1st WW domain model template, secondary structure and model quality

```
fbp21ww2      LLSKCPWKEYKSDSGKPYYY-NSQTKESRWAKP  32
f4jc80ww1     -----QWKMILHEESNQYYYWNTETGETSWELP  28
                **        :..     *** *:.* *:  *    *
```

(35.71 % sequence identity with template)
PSIPRED secondary structure prediction

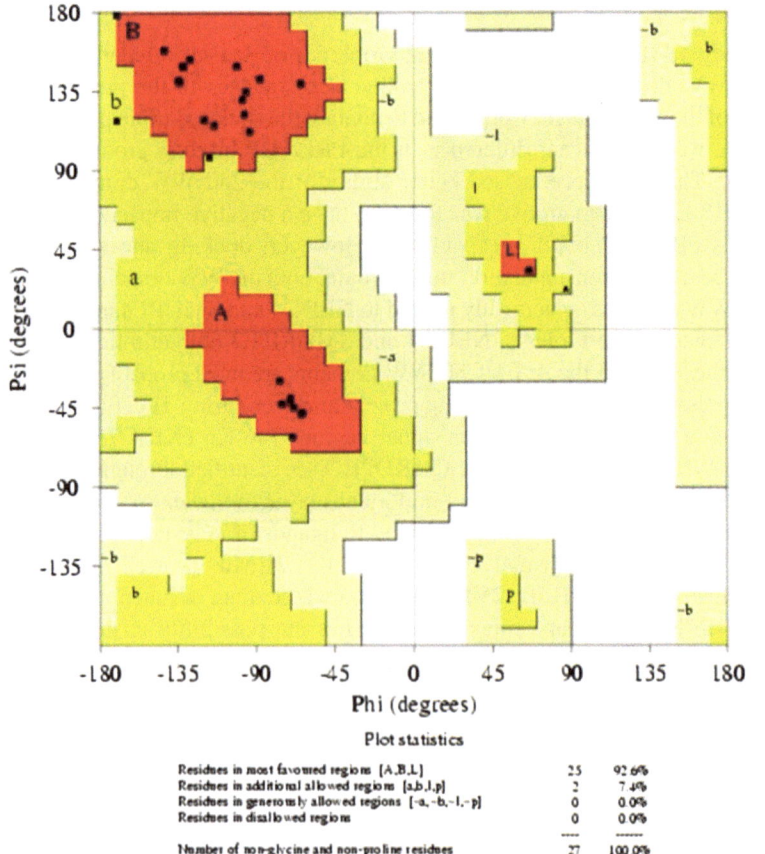

Yellow marked residues form β sheets

Plot statistics

Residues in most favoured regions [A,B,L]	25	92.6%
Residues in additional allowed regions [a,b,l,p]	2	7.4%
Residues in generously allowed regions [-a,-b,-l,-p]	0	0.0%
Residues in disallowed regions	0	0.0%
	----	------
Number of non-glycine and non-proline residues	27	100.0%

Verify_3D score: 96.77 %
Errat analysis overall quality factor: 72.77

Details of *At* F4JC80 2nd WW domain model template, secondary structure and model quality

```
sav1ww1       LPPGWSVDWTMRGRK-YYIDHNTNTTHWSHP 30
f4jc80ww2     LPSEWQAYWDESTKKVYYGNTSTSQTSWTRP 31
              **. *.. *      :* ** : .*. * *::*
```

(35.48 % sequence identity with template)
PSIPRED secondary structure prediction

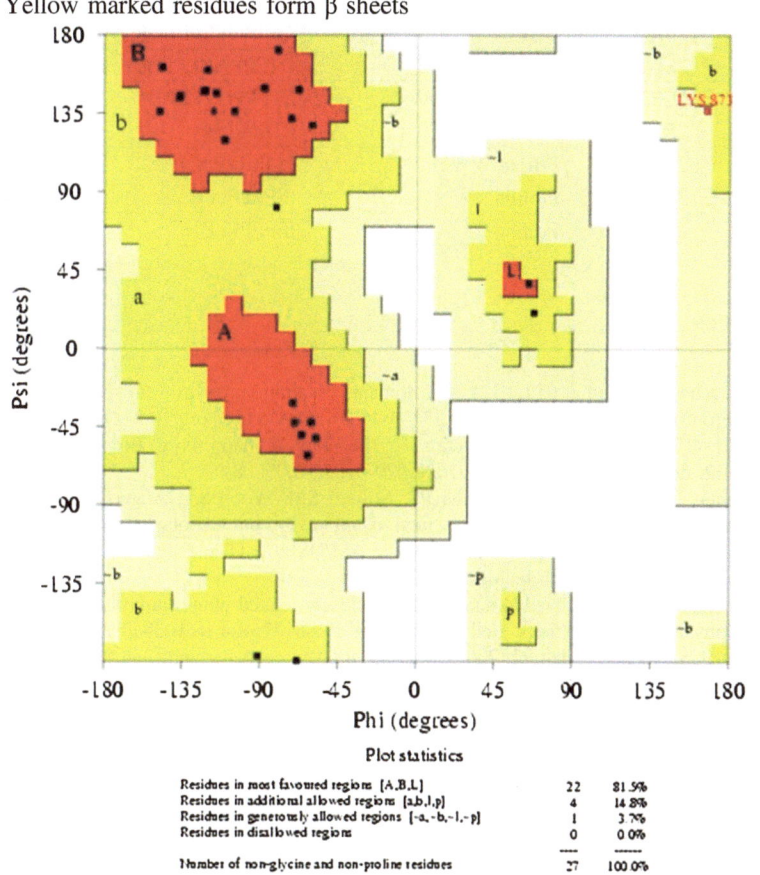

Yellow marked residues form β sheets

Plot statistics

Residues in most favoured regions [A,B,L]	22	81.5%
Residues in additional allowed regions [a,b,l,p]	4	14.8%
Residues in generously allowed regions [-a,-b,-l,-p]	1	3.7%
Residues in disallowed regions	0	0.0%
Number of non-glycine and non-proline residues	27	100.0%

Verify_3D score: 100 %
Errat analysis overall quality factor: 59.091 (Supplementary Table 1)

Supplementary Table 1 Detailed list of ligands used for docking studies from reported WW domain—ligand complex structures

Serial no.	PDB ID	PDB receptor molecule	Ligand sequence
1	1I5H	NEDD4-WW1	GSTILPIPGTPPPNYDSL
2	1JMQ	YAP1	GTPPPPYTVG
3	1K5R	YAP1	GTPPPPYTVG
4	2DJY	SMURF2-WW3	GPLGSELESPPPPYSRYPMD
5	2EZ5	NEDD4-WW3	TGLPSYDEALH
6	2HO2	FE65	PPPPPPPPPL
7	2JMF	Su(dx)-WW4	GPLGSPNTGAKQPPSYEDCIK
8	2JO9	ITCH-WW3	EEPPPPYED
9	2JUP	FBP28-WW2	PPLIPPPP
10	2KQ0	NEDD4-WW3	ILPTAPPEYMEA
11	2KXQ	SMURF2-WW2	GPLGSELESPPPPYSRYPMD
12	2LAW	YAP1	TPPPAYLPPEDP
13	2LB1	SMURF2-WW2	DTPPPAYLPPEDP
14	2LB2	NEDD4L-WW2	ETPPPGYLSEDG
15	2RLY	FBP28-WW2	PTPPPLPP
16	2RM0	FBP28-WW2	PPPLIPPPP
17	2V8F	Profilin	IPPPPPLPGV
18	2V8F	Profilin	IPPPPPLP

References

1. Salah Z, Alian A, Aqeilan RI (2012) WW domain-containing proteins: retrospectives and the future. Front Biosci (Landmark Ed) 17:331–348. doi:http://dx.doi.org/10.2741/3930
2. Sudol M (1996) Structure and function of the WW domain. Prog Biophys Mol Biol 65:113–32. doi:http://dx.doi.org/10.1016/S0079-6107(96)00008-9
3. Mouchantaf R, Azakir BA, McPherson PS, Millard SM, Wood SA, Angers A (2006) The ubiquitin ligase itch is auto-ubiquitylated in vivo and in vitro but is protected from degradation by interacting with the deubiquitylating enzyme FAM/USP9X. J Biol Chem 281:38738–38747. doi:http://dx.doi.org/10.1074/jbc.M605959200
4. Yang B, Kumar S (2010) Nedd4 and Nedd4-2: closely related ubiquitin-protein ligases with distinct physiological functions. Cell Death Differ 17:68–77. doi:10.1038/cdd.2009.84
5. DelMare S, Salah Z, Aqeilan RI (2009) WWOX: its genomics, partners, and functions. J Cell Biochem 108:737–745. doi:10.1002/jcb.22298
6. Sudol M, Shields DC, Farooq A (2012) Structures of YAP protein domains reveal promising targets for development of new cancer drugs. Semin Cell Dev Biol 23:827–833. doi:10.1016/j.semcdb.2012.05.002
7. Kato Y, Miyakawa T, Kurita J, Tanokura M (2006) Structure of FBP11 WW1-PL ligand complex reveals the mechanism of proline-rich ligand recognition by group II/III WW domains. J Biol Chem 281:40321–40329. doi:http://dx.doi.org/10.1074/jbc.M609321200
8. Lippens G, Landrieu I, Smet C (2007) Molecular mechanisms of the phospho-dependent prolyl cis/trans isomerase Pin1. FEBS J 274:5211–5222. doi:http://dx.doi.org/10.1111/j.1742-4658.2007.06057.x

9. Qin J, Barajas D, Nagy PD (2012) An inhibitory function of WW domain-containing host proteins in RNA virus replication. Virology 426:106–119. doi:http://dx.doi.org/10.1016/j. virol.2012.01.020

10. Wu R, Li S, He S, Wassmann F, Yu C, Qin G et al. (2011) CFL1, a WW domain protein, regulates cuticle development by modulating the function of HDG1, a class IV homeodomain transcription factor, in rice and Arabidopsis. Plant Cell 23:3392–3411. doi:http://dx.doi.org/ 10.1105/tpc.111.088625

11. Noh YS, Bizzell CM, Noh B, Schomburg FM, Amasino RM (2004) EARLY FLOWERING 5 acts as a floral repressor in Arabidopsis. Plant J 38:664–672

12. Hong F, Attia K, Wei C, Li K, He G, Su W et al (2007) Overexpression of the rFCA RNA recognition motif affects morphologies modifications in rice (Oryza sativa L.). Biosci Rep 27:225–234. doi:http://dx.doi.org/10.1007/s10540-007-9047-y

13. Kang CH, Feng Y, Vikram M, Jeong IS, Lee JR, Bahk JD et al (2009) *Arabidopsis thaliana* PRP40s are RNA polymerase II C-terminal domain-associating proteins. Arch Biochem Biophys 484:30–38. doi:10.1016/j.abb.2009.01.004, http://dx.doi.org/10.1016/j.abb.2009.01.004

14. Landrieu I, Wieruszeski JM, Wintjens R, Inzé D, Lippens G (2002) Solution structure of the single-domain prolyl cis/trans isomerase PIN1At from *Arabidopsis thaliana*. J Mol Biol 320:321–332. doi:http://dx.doi.org/10.1016/S0022-2836(02)00429-1

15. Punta M, Coggill PC, Eberhardt RY, Mistry J, Tate J, Boursnell C et al (2012) The Pfam protein families database. Nucleic Acids Res (database issue) 40:D290–D301. doi:http://dx. doi.org/10.1093/nar/gkr1065

16. Pieper U, Webb BM, Barkan DT, Schneidman-Duhovny D, Schlessinger A, Braberg H et al. (2011) ModBase, a database of annotated comparative protein structure models, and associated resources. Nucleic Acids Res (database issue) 39:D465–474. doi:http://dx.doi.org/ 10.1093/nar/gkq1091

17. Lüthy R, Bowie JU, Eisenberg D (1992) Assessment of protein models with three-dimensional profiles. Nature 356:83–85. doi:http://dx.doi.org/10.1038/356083a0

18. Bowie JU, Lüthy R, Eisenberg D (1991) A method to identify protein sequences that fold into a known three-dimensional structure. Science 253:164–170. doi:http://dx.doi.org/10.1126/ science.1853201

19. Colovos C, Yeates TO (1993) Verification of protein structures: patterns of nonbonded atomic interactions. Protein Sci 2:1511–1519. doi:http://dx.doi.org/10.1002/pro.5560020916

20. Wiederstein M, Sippl MJ (2007) ProSA-web: interactive web service for the recognition of errors in three-dimensional structures of proteins. Nucleic Acids Res 35:W407–410. doi:http:// dx.doi.org/10.1093/nar/gkm290

21. Holm L, Rosenström P (2010) Dali server: conservation mapping in 3D. Nucleic Acids Res 38:W545–549. doi:http://dx.doi.org/10.1093/nar/gkq366

22. Brooks BR, Bruccoleri RE, Olafson BD, States DJ, Swaminathan S, Karplus M (1983) CHARMM: a program for macromolecular energy, minimization, and dynamics calculations. J Comp Chem 4:187–217. http://dx.doi.org/10.1002/jcc.540040211

23. Pierce BG, Hourai Y, Weng Z (2011) Accelerating protein docking in ZDOCK using an advanced 3D convolution library. PLoS One 6(9):e24657. doi:http://dx.doi.org/10.1371/ journal.pone.0024657

24. London N, Raveh B, Cohen E, Fathi G, Schueler-Furman O (2011) Rosetta FlexPepDock web server—high resolution modeling of peptide-protein interactions. Nucleic Acids Res (web server issue) 39:W249–253. doi:10.1093/nar/gkr431.doi, http://dx.doi.org/10.1093/nar/gkr431

25. Nguyen, BAN, Pogoutse A, Provart N, Moses AM (2009) NLStradamus: a simple Hidden Markov model for nuclear localization signal prediction. BMC Bioinformatics 10(1):202. doi: http://dx.doi.org/10.1186/1471-2105-10-202

26. Blanchoin,L, Staiger CJ (2010) Plant formins: diverse isoforms and unique molecular mechanism. Biochem Biophys Acta 1803:201–206. doi:10.1016/j.bbamcr.2008.09.015, http:// dx.doi.org/10.1016/j.bbamcr.2008.09.015

Structural Insights into IbpA–IbpB Interactions to Predict Their Roles in Heat Shock Response

Sanchari Bhattacharjee, Rakhi Dasgupta and Angshuman Bagchi

Abstract Cells respond to stress conditions. As a result of stress, most genes are deactivated, while a few are activated with antistress response. The latter involves a variety of molecules including molecular chaperones or heat shock proteins (Shps) whose levels get increased in stressed conditions, particularly at elevated temperatures. Heat shock proteins help the other cellular proteins to achieve their native states, i.e. correct folding or functional conformations. Thus, heat shock proteins play a major role in protein homeostasis network of the cell. Small heat shock proteins (sHsps) are one of the families of molecular chaperones that prevent the irreversible aggregation and assist in the refolding of denatured proteins. Two members of the sHsp family, IbpA and IbpB, are present in *Escherichia coli*. The IbpA and IbpB proteins are 48 % identical at the amino acid sequence level and have the characteristic α-crystalline domain. It is known that the cooperation between IbpA and IbpB is crucial for their chaperone activity in heat stressed condition. So far, the molecular mechanisms of the stress response of the IbpA/IbpB protein system have not been well understood. In the present work, an attempt has been made to identify the amino acid residues of the IbpA and IbpB proteins, which are found to be involved in protein–protein interactions. The interactions between IbpA and IbpB are studied with and without the presence of substrate Lactate Dehydrogenase (LDH) at cold shock, physiological and heat shock temperatures to observe the changes in the pattern of interaction. This study is the first report to elucidate the mechanism of interactions between the proteins.

Keywords Small heat shock proteins · Heat stress · Heat shock temperature · Ibpa–IbpB interaction

S. Bhattacharjee · R. Dasgupta (✉) · A. Bagchi (✉)
Department of Biochemistry and Biophysics, University of Kalyani, Nadia 741235, West Bengal, India
e-mail: rdgadg@gmail.com

A. Bagchi
e-mail: angshuman_bagchi@yahoo.com

© The Author(s) 2015   41
N.B. Muppalaneni and V.K. Gunjan (eds.), *Computational Intelligence in Medical Informatics*, Forensic and Medical Bioinformatics,
DOI 10.1007/978-981-287-260-9_4

1 Introduction

Cells react to various types of physical (e.g. heat) or chemical (anoxia, low pH) stresses. Cell stresses are frequent and recurring which challenge the cells leading ultimately to stress response. Most of the genes are deactivated as a result of cellular stress response, but only a few become activated to combat it. The antistress response mechanism involves a variety of molecules including molecular chaperones and proteases or heat shock proteins (Shps) whose levels increase particularly at elevated temperatures [1]. Heat Shps help the other cellular proteins to achieve the native folding conformations, to localize them at their destined sub-cellular organelles, to prevent their denaturation by heat stresses and retrieve native folding states after partial denaturations. Thus, heat Shps play a vital role in cellular protein homeostasis network [2].

Small heat shock proteins (sHsps) belong to the families of molecular chaperones that assist in the refolding of denatured proteins by resisting the irreversible aggregation [3]. Small heat Shps are wide in distribution. They are characterized by their low molecular weight (12–30 kDa) and the presence of a conserved α-crystalline domain [4]. Their robust upregulation under temperature-stressed conditions make them among the most abundant cellular proteins [5].

Two members of sHsp family, IbpA and IbpB are found in *Escherichia coli*. These two proteins are 48 % identical at amino acid sequence level [6] and have the characteristic α-crystalline domain flanked by N- and C-terminal ends [7]. Initially, both are identified as inclusion body proteins but later they are found in protein aggregates in temperature-stressed cells [8]. The efficiency of IbpA/IbpB depends on increased temperature with protein reactivation and protection from degradation found at elevated temperatures [9]. It is reported that the presence of IbpA during heat denaturation of substrate is sufficient to assist to change the macroscopic properties of aggregates, yet this alone does not increase the efficiency of the subsequent reactivation of such aggregated polypeptides. The presence of IbpB is required to assist to increase the efficiency of disaggregation and refolding [10]. This observation depicts the cooperative mode of interaction of IbpA and IbpB, in which IbpA associates first with the aggregating protein or substrate and then attracts IbpB to the complex [11]. So the cooperation between IbpA and IbpB is crucial for their chaperone activity in temperature-stressed condition. So far, the molecular mechanisms of the stress response of the IbpA/IbpB protein system have not been well understood. In the present work, we tried to elucidate the interaction profiles between the proteins at the molecular level. We identified the amino acid residues of the IbpA and IbpB proteins which are found to be involved in protein–protein interactions. The interactions between IbpA and IbpB are studied with and without the presence of substrate Lactate Dehydrogenase (LDH) [12] at cold shock, physiological and heat shock temperatures to observe the changes in the pattern of interaction following heat and cold stress. This study is the first report to elucidate the mechanism of interactions between the proteins.

2 Materials and Methods

2.1 Sequence Homology Search

First, the amino acid sequences of IbpA and IbpB, comprising of 137 and 142 amino acid residues, respectively, were extracted from UniProt [13] (id:P0C054 and id:P0C058, respectively). This was then used as input sequence to search for the template in PDB [14] using PSI-BLAST [15]. In both the cases, the A chain of crystal structure of an eukaryotic small heat shock protein (PDB code:1GME) [16] was obtained as the closest match.

2.2 Homology Modelling of the Proteins

This structure (1GME, chain A) was then used as a template to build the homology models for IbpA and IbpB separately in Accelrys Discovery Studio Modeller (version 2.5) [17]. Alternatively, the models of IbpA and IbpB were also built independently using RaptorX [18] and Phyre2 [19] homology modelling servers. The models so generated were energy minimized to ensure proper interactions applying CHARMM force field [20] using conjugate gradient algorithm [21] until the structures reached the final derivative of 0.001 kcal/mol. The stereochemical qualities of the resultant structures were then checked with PROCHECK [22] and Verify3D [23] and Ramachandran plots [24] were drawn. No residues were found to be present in the disallowed regions of the Ramachandran plot.

2.3 Molecular Docking Simulations with IbpA and IbpB

The model of IbpA was docked with the model of IbpB using Zdock [25], PyDock [26], Patchdock [27], GrammX [28] and Cluspro [29] servers in order to get a comprehensive result. Docked complexes with best score were selected and the stereochemical qualities of the models were checked with PROCHECK and Verify3D. The docked complex with the best PROCHECK and Verify3D score was selected as the final working model. The final docked complex was then energy minimized to ensure proper interactions. 918 cycles of steepest descent (SD) energy minimization [30] were performed in explicit solvent system [31] applying CHARMM force field in GROMACS version 4.6.5 [32] until the structure reached the final derivative of 0.001 kcal/mol. The stereochemical qualities of minimized docked complex were again tested with PROCHECK and Verify3D.

Next, the crystal structure of the substrate, i.e. LDH was retrieved from PDB (PDB code-3H3F) [33] and the stereochemical qualities were tested with PRO-CHECK and Verify3D. Then, this structure was again docked with docked complex of IbpAB using Zdock, PyDock, Patchdock, GrammX and Cluspro. The stereochemical qualities of the docked complexes so generated were tested with PRO-CHECK and Verify3D. Docked complex with the best score was selected as the final working model and was then energy minimized to ensure proper interactions using SD algorithm in explicit solvent system applying CHARMM force field in GROMACS version 4.6.5 until the structure reached the final derivative of 0.001 kcal/mol. The docked complex was minimized in 875 steps using SD algorithm.

2.4 Molecular Dynamics Simulation of the Complexes

In order to study the stress response both the minimized docked complexes (i.e. IbpA–IbpB and IbpAB-LDH) were then elevated from cold shock temperature (300 K) to physiological (310 K) and heat shock temperature (318 K) [34] separately and 20 ns of production runs were performed in GROMACS version 4.6.5 in each of the cases, keeping pressure constant. The conformational changes were recorded and plotted against time. Further analyses were done using Discovery Studio Package version 2.5.

3 Results

Analyses of docked complexes have revealed that, irrespective of the presence or absence of substrate, IbpA and IbpB have followed a definite pattern of interactions following cold shock and heat shock temperature in contrast to physiological temperature. It has been observed that, in case of IbpA–IbpB docked complex, 12 amino acid residues are involved in forming H-bond interactions in cold shock temperature (300 K) (Fig. 1). The interactions go down at the physiological temperature of (310 K) which involve only 10 amino acid residues (Fig. 2). However, with rise in temperature from 300 to 310 K the previous 12 amino acid residues which were involved in H-bond interactions in cold shock temperature are restored when the docked complex is elevated to heat shock temperature (318 K) (Fig. 3) (Table 1).

In presence of the substrate LDH, it has been observed that the IbpAB-LDH docked complex has 18 amino acid residues forming H-bond interactions in cold shock temperature, among which 8 amino acid residues from IbpA/B interact with the substrate LDH and the remaining 10 residues show inter-protein H-bond

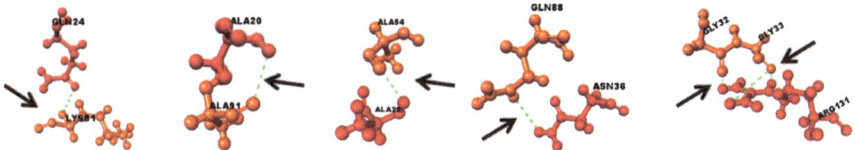

Fig. 1 Amino acid residues from IbpA and IbpB proteins involved in intermolecular H-bond interaction in IbpAB protein complex in absence of substrate at cold shock temperature (300 K) [backbone of IbpA is marked in *orange* and backbone of IbpB is marked in *red*. H-bonds are marked in *green dashed lines*] (the relevant interacting residues that are unique to this conformation are shown)

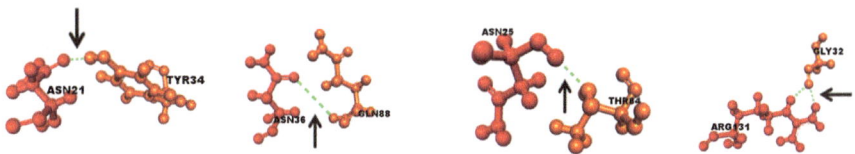

Fig. 2 Amino acid residues from IbpA and IbpB proteins involved in intermolecular H-bond interaction in IbpAB protein complex in absence of substrate at physiological temperature (310 K) [backbone of IbpA is marked in *orange* and backbone of IbpB is marked in *red*. H-bonds are marked in *green dashed lines*] (the relevant interacting residues that are unique to this conformation are shown)

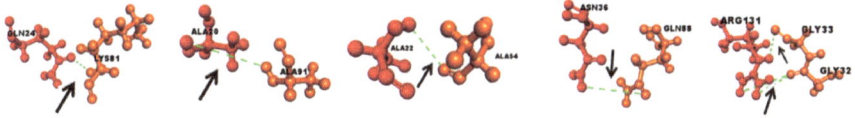

Fig. 3 Amino acid residues from IbpA and IbpB proteins involved in intermolecular H-bond interaction in IbpAB protein complex in absence of substrate at heat shock temperature (318 K) [backbone of IbpA is marked in *orange* and backbone of IbpB is marked in *red*. H-bonds are marked in *green dashed lines*] (the relevant interacting residues that are unique to this conformation are shown)

interactions (Fig. 4). When this docked complex (IbpAB with substrate LDH) is brought to physiological temperature, only 15 amino acid residues are involved in H-bond interactions. Of these 15 amino acid residues, 6 residues interact with LDH and the rest of them show inter-protein interactions (Fig. 5). When the complex again is elevated to heat shock temperature, the docked complex restores the interaction pattern as it shows 18 amino acid residues participating in H-bond interactions, among which 8 amino acid residues interact with the substrate LDH and the remaining 10 amino acid residues have inter-protein interactions (Fig. 6) (Table 2).

<...></...>

Table 1 Amino acid residues from IbpA and IbpB proteins involved in interactions in the IbpAB protein complex in the absence of substrate at cold shock, physiological and heat shock temperatures [interacting residues that remain unchanged at these three temperatures are marked in bold pattern. IbpA is denoted as A, IbpB is denoted as B]

300 K Donor	300 K Acceptor	310 K Donor	310 K Acceptor	318 K Donor	318 K Acceptor
A:LYS81:HN	B:GLN24:OE1	A:TYR34:HH	B:ASN21:OD1	A:LYS81:HN	B:GLN24:OE1
A:GLN88:HN	**B:TYR35:O**	**A:GLN88:HN**	**B:TYR35:O**	**A:GLN88:HN**	**B:TYR35:O**
B:ALA20:HN	A:ALA91:O	A:GLN88:HE21	B:ASN36:OD1	B:ALA20:HN	A:ALA91:O
B:ALA22:HN	A:ALA54:O	**B:ALA22:HN**	**A:GLU92:OE2**	B:ALA22:HN	A:ALA54:O
B:ALA22:HN	**A:GLU92:OE2**	**B:GLN24:HE22**	**A:GLU82:O**	**B:ALA22:HN**	**A:GLU92:OE2**
B:GLN24:HE22	**A:GLU82:O**	B:ASN25:HN	A:THR84:OG1	**B:GLN24:HE22**	**A:GLU82:O**
B:ASN25:HD21	**A:GLU82:OE2**	**B:ASN25:HD21**	**A:GLU82:OE2**	**B:ASN25:HD21**	**A:GLU82:OE2**
B:TYR35:HN	**A:TYR85:O**	**B:TYR35:HN**	**A:LEU86:O**	**B:TYR35:HN**	**A:TYR85:O**
B:ASN36:HD22	A:GLN88:OE1	B:ARG131:HE	A:GLY32:O	B:ASN36:HD22	A:GLN88:OE1
B:ARG131:HE	A:GLY33:O	B:ARG131:HH21	A:GLY32:O	B:ARG131:HE	A:GLY33:O
B:ARG131:HH12	A:GLY32:O			B:ARG131:HH12	A:GLY32:O
B:ARG131:HH22	A:GLY32:O			B:ARG131:HH22	A:GLY32:O

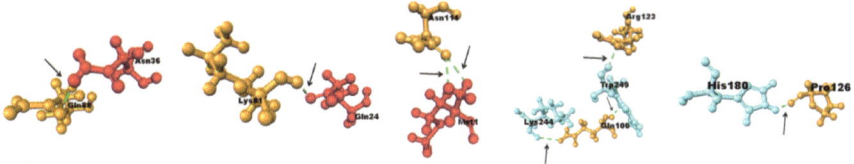

Fig. 4 Amino acid residues from IbpA and IbpB proteins involved in intermolecular H-bond interaction in IbpAB protein complex in presence of substrate LDH at cold shock temperature (300 K) [backbone of IbpA is marked in *orange*, backbone of IbpB is marked in *red* and backbone of substrate LDH is marked in *blue*. H-bonds are marked in *green dashed lines*] (the relevant interacting residues that are unique to this conformation are shown)

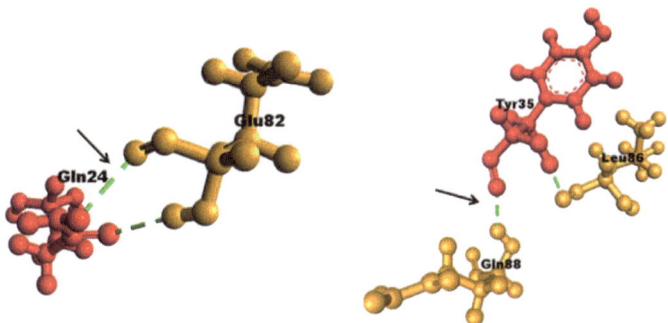

Fig. 5 Amino acid residues from IbpA and IbpB proteins involved in intermolecular H-bond interaction in IbpAB protein complex in the presence of substrate at physiological temperature (310 K) [backbone of IbpA is marked in *orange*, backbone of IbpB is marked in *red* and backbone of substrate LDH is marked in *blue*. H-bonds are marked in *green dashed lines*] (the relevant interacting residues that are unique to this conformation are shown)

4 Discussions

Escherichia coli small heat Shps, IbpA and IbpB, function as molecular chaperones and protect misfolded proteins against irreversible aggregation as well as help the unfolded proteins to restore their native conformations following heat stress.

In our study, it is observed that irrespective of presence or absence of substrate LDH, the number of interactions increases both in heat shock and cold shock temperatures in comparison to physiological temperature. In presence of substrate LDH, at cold shock temperature (300 K) 8 amino acid residues of IbpA and IbpB interact with substrate LDH and 10 residues show inter-protein interactions. Among these 8

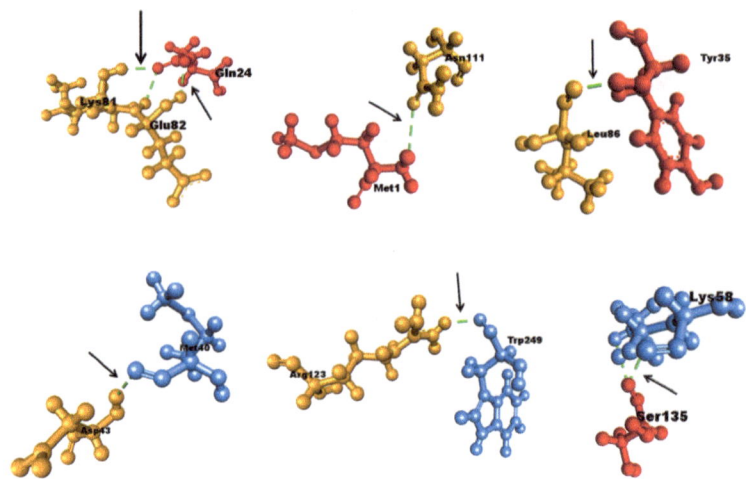

Fig. 6 Amino acid residues from IbpA and IbpB proteins involved in intermolecular H-bond interaction in IbpAB protein complex in the presence of substrate at heat shock temperature (318 K) [backbone of IbpA is marked in *orange*, backbone of IbpB is marked in *red* and backbone of substrate LDH is marked in *blue*. H-bonds are marked in green dashed lines] (the relevant interacting residues that are unique to this conformation are shown)

residues, IbpA is involved in seven interactions, whereas the remaining one is contributed by IbpB. At physiological temperature (310 K), only two amino acid residues of IbpB interact with the substrate, whereas four amino acid residues of IbpA interact with the same and only 9 residues show inter-protein interactions. When the complex is brought to heat shock temperature (318 K), 8 amino acid residues of IbpAB interact with substrate LDH, of which five residues are contributed by IbpA and remaining three residues by IbpB, whereas at heat shock temperature, another 10 residues are involved in inter-protein interactions. In each of the cases, the number of interacting residues is more in IbpA than IbpB, which suggests that IbpA interacts with the substrate first, thus allowing it to be more exposed to interact with IbpB to make efficient refolding of heat stressed substrate. It is also noted that inter-protein interaction increases at cold shock and heat shock temperature in comparison to physiological temperature, which indicates that in heat stressed condition, the cooperative interaction between IbpA and IbpB is also important for their chaperone activity. So far, this is the first study to elucidate the molecular mechanism of the binding interactions of IbpA and IbpB at heat stress conditions. This work would therefore be essential to predict the hitherto unknown molecular mechanisms of IbpAB interactions.

Table 2 Amino acid residues from IbpA and IbpB proteins involved in interactions in the IbpAB protein complex in the presence of substrate at cold shock, physiological and heat shock temperature

300 K		310 K		318 K	
Donor	Acceptor	Donor	Acceptor	Donor	Acceptor
C:HIS66:HE2	**B:ALA100:O**	**C:LYS58:HZ1**	**B:SER135:O**	C:LYS58:HZ1	B:SER135:O
C:HIS180:HE2	A:PRO126:O	**C:HIS66:HE2**	**B:ALA100:O**	C:LYS58:HZ3	B:SER135:O
C:LYS244:HZ2	**A:LEU101:O**	**C:LYS244:HZ2**	**A:LEU101:O**	**C:HIS66:HE2**	**B:ALA100:O**
C:TYR246:HH	**A:GLN100:OE1**	**C:TYR246:HH**	**A:GLN100:OE1**	**C:LYS244:HZ2**	**A:LEU101:O**
C:TRP249:HE1	A:GLN100:O	A:ASN38:HD22	C:ASP63:O	**C:TYR246:HH**	**A:GLN100:OE1**
A:ARG48:HH22	**C:PHE70:O**	**A:ARG48:HH22**	**C:PHE70:O**	A:ASP43:HN	C:MET40:O
A:LYS81:HN	B:GLN24:OE1	A:GLU82:HN	B:GLN24:OE1	**A:ARG48:HH22**	**C:PHE70:O**
A:GLN100:HE22	C:LYS244:O	A:GLN88:HN	B:TYR35:O	A:LYS81:HN	B:GLN24:OE1
A:ARG123:HH12	C:TRP249:O	**B:ASN21:HD21**	**A:TYR34:OH**	A:GLU82:HN	B:GLN24:OE1
B:MET1:H1	A:ASN111:OD1	**B:ALA22:HN**	**A:GLU92:OE2**	A:ARG123:HH12	C:TRP249:O
B:MET1:H3	A:ASN111:OD1	**B:GLN24:HE22**	**A:GLU82:O**	B:MET1:H2	A:ASN111:OD1
B:ASN21:HD21	**A:TYR34:OH**	**B:ASN25:HN**	**A:THR84:OG1**	**B:ASN21:HD22**	**A:TYR34:OH**
B:ALA22:HN	**A:GLU92:OE2**	B:TYR35:HN	A:LEU86:O	**B:ALA22:HN**	**A:GLU92:OE2**
B:GLN24:HE22	**A:GLU82:O**	**B:ARG131:HE**	**A:GLY32:O**	**B:GLN24:HE22**	**A:GLU82:O**
B:ASN25:HN	**A:THR84:OG1**	**B:ARG131:HH21**	**A:GLY32:O**	**B:ASN25:HN**	**A:THR84:OG1**
B:ASN36:HD22	A:GLN88:OE1			B:TYR35:HN	A:LEU86:O
B:ARG131:HE	**A:GLY32:O**			**B:ARG131:HE**	**A:GLY32:O**
B:ARG131:HH21	**A:GLY32:O**			**B:ARG131:HH21**	**A:GLY32:O**

Interacting residues that remain unchanged at these three temperatures are marked in bold pattern. Amino acid residues from IbpA and IbpB of IbpAB protein complex that interact with the substrate are marked in blue. IbpA is denoted as A, IbpB is denoted as B and substrate is denoted as C

Acknowledgments The authors are thankful to Department of Biochemistry and Biophysics and BIF centre, University of Kalyani for their continuous support and for providing the necessary instruments to carry out the experiments. The authors would like to acknowledge the ongoing DST-PURSE programme (2012–2015) and DBT (project no. BT/PR6869/BID/7/417/2012) for support.

Conflict of Interest
The authors declare no conflict of interests.

References

1. Gill RT, Valdes JJ, Bentley WE (2000) A comparative study on global stress gene regulation in response to over-expression of recombinant proteins in *Escherichia coli*. Metab Eng 2:178–189
2. Thomas JG, Baneyx F (1998) Roles of the *Escherichia coli* small heat shock proteins IbpA and IbpB in thermal stress management: comparison with ClpA, ClpB, and HtpG in vivo. J Bacteriol 180:5165–5172
3. Kitagawa M, Matsumura Y, Tsuchido T (2000) Small heat shock proteins, IbpA and IbpB, are involved in resistances to heat and superoxide stresses in *Escherichia coli*. FEMS Microbiol Lett 184:165–171
4. Kuczyńska-Winik D, Kdzierska S, Matuszewska E, Lund P, Taylor A, Lipinska B, Laskowska E (2002) The *Escherichia coli* small heat shock proteins IbpA and IbpB prevent the aggregation of endogenous proteins denatured in vivo during extreme heat shock. Microbiology 148:1757–1765
5. Carrió MM, Villaverde A (2003) Role of molecular chaperones in inclusion body formation. FEBS Lett 537:215–221
6. Strózecka J, Chrusciel E, Górna E, Szymanska A, Zietkiewicz S, Liberek K (2012) Importance of N- and C-terminal regions of IbpA, *Escherichia coli* small heat shock protein, for chaperone function and oligomerization. J Biol Chem 287:2843–2853. doi:10.1074/jbc.M111.273847
7. Van Montfort R, Slingsby C, Vierling E (2001) Structure and function of the small heat shock protein α-crystallin family of molecular chaperones. Adv Protein Chem 59:105–156
8. Allen SP, Polazzi JO, Gierse JK, Easton AM (1992) Two novel heat shock genes encoding proteins produced in response to heterologous protein expression in *Escherichia coli*. J Bacteriol 174:6938–6947
9. Lee GJ, Roseman AM, Saibil HR, Vierling E (1997) A small heat shock protein stably binds heat-denatured model substrates and can maintain a substrate in a folding-competent state. EMBO J 16:659–671
10. Jiao W, Qian M, Li P, Zhao L, Chang Z (2005) The essential role of the flexible termini in the temperature-responsiveness of the oligomeric state and chaperone-like activity for the polydisperse small heat shock protein IbpB from *Escherichia coli*. J Mol Biol 347:871–884
11. Kuczyńska-Winik D, Kedzierska S, Matuszewska E, Lund P, Taylor A, Lipińska B, Laskowska E (2002) The *Escherichia coli* small heat shock proteins IbpA and IbpB prevent the aggregation of endogenous proteins denatured *in vivo* during extreme heat shock. Microbiology 148:1757–1765
12. Motohashi K, Watanabe Y, Yohda M, Yoshida M (1999) Heat-inactivated proteins are rescued by the DnaKJ-GrpE set and ClpB chaperones. Pnas 13:7184–7189. doi:10.1073/pnas.96.13.7184
13. Leinonen R, Diez FG, Binns D, Fleischmann W, Lopez R, Apweiler R (2004) UniProt archive. Bioinformatics 20(17):3236–3237. doi:10.1093/bioinformatics/bth191 PMID15044231
14. Berman HM (2008) The Protein Data Bank: a historical perspective. Acta Crystallogr Sect A Found Crystallogr A64(1):8895. doi:10.1107/S0108767307035623, PMID 18156675

15. Camacho C, Coulouris G, Avagyan V, Ma N, Papadopoulos J, Bealer K, Madden TL (2009) BLAST+: architecture and applications. BMC Bioinformatics 10:421. doi:10.1186/1471-2105-10-421, PMID 20003500

16. Van Montfort RLM, Basha E, Friedrich KL, Slingsby C, Vierling E (2001) Crystal structure and assembly of a eukaryotic small heat shock protein. Nat Struct Biol 8:1025. doi:10.1038/nsb722, PubMed: 11702068

17. Eswar N, Marti-Renom MA,Webb B, Madhusudhan MS, Eramian D, Shen M, Pieper U, Sali A (2006) Comparative protein structure modeling with MODELLER. In: Current protocols in bioinformatics, vol 15, Wiley, New York, pp 5.6.1–5.6.30

18. Peng J, Jinbo X, Raptor X (2011) Exploiting structure information for protein alignment by statistical inference. Proteins 79:161–71. doi:10.1002/prot.23175, PMC 3226909, PMID 21987485

19. Bennett-Lovsey RM, Herbert AD, Sternberg MJE, Kelley LA (2007) Exploring the extremes of sequence/structure space with ensemble fold recognition in the program Phyre. Proteins Struct Funct Bioinf 70(3):611. doi:10.1002/prot.21688

20. Brooks BR, Bruccoleri RE, Olafson BD, States DJ, Swaminathan S, Karplus M (1983) CHARMM: a program for macromolecular energy, minimization, and dynamics calculations. J Comp Chem 4(2):187–217. doi:10.1002/jcc.540040211

21. Fletcher R, Powell MJD (1963) A rapidly convergent descent method for minimization. Comput J 6:163–168

22. Laskowski RA, MacArthur MW, Moss DS, Thornton JM (1993) PROCHECK: a program to check the stereochemical quality of protein structures. J Appl Cryst 26:283–291

23. Lüthy R, Eisenberg JU, Bowie D (1992) Assessment of protein models with three-dimensional profiles. Nature 356(6364):83–85

24. Ramachandran GN, Ramachandran C, Sasisekharan V (1963) Stereochemistry of polypeptide chain configurations. J Mol Biol 7:95–99. doi:10.1016/S0022-2836(63)80023-6 PMID 13990617

25. Chen R, Weng Z (2003) ZDOCK: an initial-stage protein-docking algorithm. Proteins 52:80–87

26. Jimenez-Garcia B, Pons C, Fernandez-Recio J (2013) pyDockWEB: a web server for rigid body protein-protein docking using electrostatics and desolvation scoring. Bioinformatics 29 (13):1698–1699

27. Schneidman-Duhovny D, Inbar Y, Nussinov R, Wolfson HJ (2005) PatchDock and SymmDock: servers for rigid and symmetric docking. Nucleic Acids Res 33:W363–W367

28. Tovchigrechko A, Vakser IA (2006) GRAMM-X public web server for protein–protein docking. Nucleic Acids Res 34:W310–W314

29. Comeau SR, Gatchell DW, Vajda S, Camacho CJ (2004) ClusPro: a fully automated algorithm for protein-protein docking. Nucleic Acids Res 32(Web Server issue):W96-9. PMID 15215358

30. Deift P, Zhou X (1993) A steepest descent method for oscillatory Riemann-Hilbert problems. Asymptotics for the MKdV equation. Ann Math 137(2):295–368. doi:10.2307/2946540

31. Lee MS, Salsbury FR, Olson MA (2004) An efficient hybrid explicit/implicit solvent method for biomolecular simulations. J Comput Chem 25(16):1967–1978. doi:10.1002/jcc.20119 PMID 15470756

32. Hess B, Kutzner C, Van Der Spoel D, Lindahl E (2008) GROMACS 4: algorithms for highly efficient, load-balanced, and scalable molecular simulation. J Chem Theory Comput 4 (2):435–447. doi:10.1021/ct700301

33. Swiderek K, Panczakiewicz A, Bujacz A, Bujacz G, Paneth P (2009) Modelling of isotope effects on binding oxamate to lactic dehydrogenase. J Phys Chem B 113(38):12782–12789. doi:10.1021/jp903579

34. Spiess C, Beil A, Ehrmann M (1999) A temperature-dependent switch from chaperone to protease in a widely conserved heat shock protein. Cell 97(3):339–347

Improving the Performance of Multi-parameter Patient Monitor System Using Additional Features

S. Premanand, C. Santhosh Kumar and A. Anand Kumar

Abstract Multi-parameter patient monitor (MPM) keep track of the condition of a patient in intensive care units (ICU) or general wards using the human vital parameters, heart rate, blood pressure, respiration rate and oxygen saturation (SpO_2). A high accuracy for the overall classification, specificity and sensitivity is extremely important in providing quality health care to the patients. Support vector machine (SVM) is a powerful supervised algorithm that is effectively used in MPMs for classification. A careful study of the vital parameters in a healthy person reveals that there exists an intrinsic relationship between the four vital parameters, for example when heart rate is on the higher side, blood pressure is expected to be on the lower side and vice versa. Hence, it would be highly required to understand the correlation between the vital parameters and to integrate it into the MPM system. In this work, we present the results of the MPM using the SVM as back-end classifier. Further, we use correlation features (feature expansion) along with base parameters in an effort to improve the performance of MPM and note that the performance of the MPM enhanced significantly.

Keywords Multi-parameter patient monitor · SVM · Vital parameters · Feature expansion · Intersection kernel

1 Introduction

Multi-parameter patient monitoring (MPM) plays an important role in ensuring quality health care in the intensive care units (ICU) and general wards to continuously monitor patients' vital parameters, heart rate, blood pressure, respiratory rate

S. Premanand · C. Santhosh Kumar
Machine Intelligence Research Laboratory, Department of Electronics
and Communication Engineering, Amrita Vishwa Vidyapeetham, Ettimadai 641112, India

A. Anand Kumar (✉)
Department of Neurology, Amrita Institute of Medical Sciences, Cochin 682041, India
e-mail: newgen.anand@gmail.com

© The Author(s) 2015   53
N.B. Muppalaneni and V.K. Gunjan (eds.), *Computational Intelligence
in Medical Informatics*, Forensic and Medical Bioinformatics,
DOI 10.1007/978-981-287-260-9_5

and oxygen saturation (SpO$_2$) and alert as and when the condition of the patient deteriorates. Studies of vital parameters to show evidence of physiological deterioration even before the patient's abnormal condition, which leads to improvement in mortality rates [1]. The current early warning score (EWS) [2] systems assign scores to each vital sign based on clinical experience, depending on its diversion away from some assumed normal range. If the scores for a vital parameters, exceeds threshold, then a clinical analysis of the patient is prompted. There is a substantial error rate associated with manual scoring and they do not consider the intrinsic relationship between the vital parameters that exist in healthy people [3–6].

Machine learning techniques can be applied to detect physiological deterioration in patients' health, with high accuracy as compared to the EWS system [7]. MPM uses support vector machine (SVM) classification, found to be effective in providing state-of-the-art performance [8, 9]. The vital parameter data gathered from the bedside monitors can be utilized to train an MPM system that could be used to effectively predict the condition of a patient under observation. One class SVMs [10] using the prior intelligence gathered from some assumed "normal" behaviour was found to be effective in the development of MPMs. However, when sufficient samples from the "abnormal" condition is available, it was seen that the two class SVMs [11] outperform one class SVMs.

Studies on the vital parameters in a healthy person suggests that there exists a well-established intrinsic relationship [3–5] between the four vital parameters, for example, when heart rate tend to increase, blood pressure is expected to get lower and vice versa.

In this work, we capture the correlation between human vital parameters, to capture the intrinsic relationship between the parameters for diagnosis more accurately and also achieve higher sensitivity, specificity and overall classification accuracy. We experimented with SVM algorithm [8] using linear, non-linear and homogeneous kernels for constructing a model to verify the effectiveness of the proposed approach and to enhance the operation of the MPMs. The system developed in the present work uses MIMIC-II database to determine the developments in the vital parameters.

2 Database Used for the Experiments

Experiments were performed on the vital parameters, obtained from MIMIC-II database [12, 13]. MIMIC-II consists of two major parts, namely clinical data and physiological waveforms. From bedside patient monitors (Component Monitoring System Intellivue MP-70; Philips Healthcare), the physiological waveforms are collected, used in medical, surgical, coronal care and neonatal ICUs in a tertiary hospital. The waveform database includes high resolution (125 Hz) waveforms (e.g., ECG), derived numeric time series such as heart rate, blood pressure, respiration rate and oxygen saturation [12]. The database also consists of monitor generated alarms along with the physician annotated files.

3 Support Vector Machine

Support Vector Machine [8] is a set of supervised learning methods used for classification and regression [14], given by a set of data with associated labels. SVM constructs a hyperplane that separates two classes, and the separating hyperplane is chosen to maximize the margin between the examples of two classes. Breaking up the classes with a big margin, minimizes a bound on expected generalization error (chance of predicting error during classification). On the bounding planes, the points in the data set which are falling are called support vectors, sv. If the given data points are non-linear kernel, $k(x, y)$, where x and y are the points in the dataset with labels naming, l_x and l_y, with $l_x, l_y \in \pm 1$. Classification of the test vector x is defined by,

$$l_x = \text{sign} \left(\sum_{i=1}^{M} \alpha_i k(x, sv(i)) \right) \tag{1}$$

where $sv(i)$ means ith support vector, M is the number of support vectors and α be the dual representation of separating hyperplane's normal vector.

In this work, we experiment with basic kernels, linear, polynomial and RBF kernels and the additive kernels [15] like intersection [16, 17], Chi-square [15] and JS kernels [18], in the SVM backend used in the MPM system.

1. Linear kernel: $K_{\text{Linear}}(x, y) = x^T y$
2. Polynomial kernel: $K_{\text{Polynomial}}(x, y) = (1 + x^T y)^d$ for any d > 0
3. RBF kernel: $K_{\text{RBF}}(x, y) = \exp\left(\frac{-1\|x-y\|^z}{2\sigma^2} \right)$
4. Intersection kernel: $K_{\text{Intersection}}(x, y) = \Sigma_{i=1}^{N} \min(x_i, y_i)$
5. Chi-square kernel: $K_{\text{Chi-square}}(x, y) = \Sigma_{i=1}^{N} \frac{x_i.y_i}{x_i+y_i}$
6. JS kernel: $K_{\text{JS}}(x, y) = \Sigma_{i=1}^{N} \frac{x_i}{2} \log_2 \frac{(x_i+y_i)}{x_i} + \frac{y_i}{2} \log_2 \frac{(x_i+y_i)}{y_i}$

The kernel, $K(x, y)$, interpreted as a measure of similarity [19] between the two examples, x and y. N is the dimension of the feature vector.

4 Improving the Performance of Multi-parameter Patient Monitor

The rejection of additional algorithmic complexity, towards the improving the performance of the MPM, we use geometric means of the vital parameters taken in pairs of two, in addition to four vital parameters used in the baseline system. Additional information about the balance/imbalance of the patients' health status are there in the expanded feature set, for example, the relationship between the parameters exposes the risk factor for problems like cardiovascular abnormalities, hyperoxia (high oxygen content in the blood), hypoxia (low oxygen content in the blood), etc. The baseline system features [6] set is given by,

$$F = [x1, x2, x3, x4, \sqrt{x1x2}, \sqrt{x1x3}, \sqrt{x1x4}, \sqrt{x2x3}, \sqrt{x2x4}, \sqrt{x3x4}] \quad (2)$$

In this work, we gathered more additional information by increasing the feature set to G, to make the performance of MPM improved, and the feature set G is given by,

$$G = \left[x1, x2, x3, x4, \sqrt{x1x2}, \sqrt{x1x3}, \sqrt{x1x4}, \sqrt{x2x3}, \sqrt{x2x4}, \sqrt{x3x4}, \right.$$
$$\left. \frac{x1x2}{x1+x2}, \frac{x1x3}{x1+x3}, \frac{x1x4}{x1+x4}, \frac{x2x3}{x2+x3}, \frac{x2x4}{x2+x4}, \frac{x3x4}{x3+x4} \right] \quad (3)$$

where $x1$, $x2$, $x3$ and $x4$ are the human vital parameters, heart rate, blood pressure, respiration rate and oxygen saturation (SpO_2) values, respectively, and the remaining correlation features helps to capture the intrinsic relationship between the vital parameters along with the help of explicit feature maps for addictive kernels such as intersection, Chi square and JS kernels.

5 Experimental Results

The MIMIC-II database consists of four vital parameters collected from 413 patients. Among them, data from 12 patients were not suitable, and the remaining 401 patients' data are used for the experiment. From the 401 patients, 1454,010 samples are separated as, 1100,510 (300 patients) samples as training data and 311,423 (101 patients) samples as testing data. Afterwards, the integral training and testing data are shuffled randomly as 50,000 and 20,000 samples with the corresponding labels. The process is repeated to generate samples for seven independent trials, and the results obtained from these trails are averaged to obtain the final result. We used LIBSVM for all our experiments. Table 1 illustrates the performance of SVM MPM system by using four vital parameters, Table 2 illustrates the performance of SVM MPM system by using four vital parameters with additional six correlation features by using geometric mean calculation thereby, increasing the efficiency and Table 3 explains the expansion of six features with the previous process by the proportion of geometric mean to the arithmetic mean calculation, hence we achieve better efficiency than by 10-feature parameter. Results conclude that the use of correlation features (16 features) along with the four vital parameters helped to enhance the MPM system performance when compared to the baseline system with ten features.

Table 1 Result for 4 feature by using baseline kernels along with homogeneous kernels

Kernels	Overall accuracy	Sensitivity	Specificity
Linear	77.14	1.55	100
Polynomial	92.62	79.54	96.58
RBF	97.38	95.54	97.93
Intersection	99.61	98.36	99.99
Chi-squared	93.56	79.65	97.77
JS	92.76	77.59	97.35

Table 2 Result for 10 feature using baseline kernels along with homogeneous kernels

Kernels	Overall accuracy	Sensitivity	Specificity
Linear	91.94	77.09	96.43
Polynomial	95.06	86.48	97.72
RBF	96.82	95.86	97.67
Intersection	99.02	97.84	99.56
Chi-squared	93.08	80.09	97.16
JS	91.92	77.48	96.37

Table 3 Result for 16 feature using baseline kernels along with homogeneous kernels

Kernels	Overall accuracy	Sensitivity	Specificity
Linear	92.61	79.44	96.60
Polynomial	95.19	86.57	97.76
RBF	97.91	96.25	98.41
Intersection	99.87	98.39	99.97
Chi-squared	93.84	80.84	97.82
JS	92.95	78.61	97.29

6 Conclusion

We notice that the values of the four vital parameters heart rate, blood pressure, respiration rate and oxygen saturation are always positive quantities, and thus the correlation features derived from these four vital parameters. In this report, we proposed a novel approach to improve the process of the MPMs taking advantage of the intrinsic relationship between the vital parameters, using additional six features by a proportion of geometric mean to the arithmetic mean calculation of the vital parameters required in a pair of two, making the total number of features in the proposed system as 16. In our experiments with the feature expansion for MPM, we are able to make an improvement in the proposed system. We measured the baseline system and the proposed system using sensitivity, specificity and overall classification accuracy.

References

1. National Patient Safety Association and others (2007) Safer care for acutely ill patients: learning from serious accidents. Technical report, NPSA
2. Tarassenko L, Clifton DA, Pinsky MR, Hravnak MT, Woods JR, Watkinson PJ (2011) Centile-based early warning scores derived from statistical distributions of vital signs. Resuscitation 82(8):1013–1018
3. Dornhost AC, Howard P (1952) Respiratory variations in blood pressure. Circulation 6:553–558
4. Yasuma F, Hayano J (2004) Respiratory sinus arrhythmia: why does the heart beat synchronize with respiratory rhythm? Chest 2:683–690
5. Beata G, Anna S et al (2013) Relationship between heart rate variability, blood pressure and arterial wall properties during air and oxygen breathing in healthy subjects. Auton Neurosci 178(1–2):60–66
6. Vishnuprasad K, Santhosh Kumar C, Ramachandran KI, Vaijeyanthi V, Anand Kumar A (2014) Towards building low cost multi-parameter patient monitors, ICC, 10 Apr 2014
7. Chan AB, Vasconcelos N, Moreno PJ (2004) A family of probabilistic kernels based on information divergence, Statistical Visual Computing Laboratory, SVCL-TR 2004/01, June 2004
8. Vapnik VN (1999) The nature of statistical learning theory, 2nd edn. Springer, Berlin, pp 237–240, 263–265, 291–299
9. Chang C-C, Lin C-J (2013) LIBSVM: a library for support vector machines. Initial version: 2001 last updated: March 2013, pp 553–558
10. Clifton L, Clifton DA, Watkinson PJ, Tarassenko (2011) Identification of patient deterioration in vital-sign data using one-class support vector machines. In: IEEE federated conference computer science and information systems (FedCSIS), pp 125–131
11. Khalid S, Clifton DA, Clifton L, Tarassenko L (2012) A two-class approach to the detection of physiological deterioration in patient vital signs, with clinical label refinement. IEEE Trans Inf Technol Biomed 16(6):1231
12. Lee J, Scott DJ et al (2011) Open-access MIMIC-II database for intensive care research. In: 33rd annual international conference of the IEEE EMBS
13. MIMIC II Database. www.physionet.org/physiobank/database/mimic2db/
14. Ruiz-Llata M, Guarnizo G, Yebenes-Calvino M (2010) FPGA implementation of a support vector machine for classification and regression. In: The 2010 international joint conference on IEEE neural networks (IJCNN), pp 1–5
15. Vedaldi A, Zisserman A (2011) Efficient additive kernels via explicit features maps. IEEE Trans Pattern Anal Mach Intell, 1–14
16. Barla A, Odone F, Verri A (2003) Histogram intersection kernel for image classification. In: Proceeding of ICIP
17. Sharma G, Jurie F (2013) A novel approach for efficient SVM classification with histogram intersection kernel. Oral presentation at the British Machine Vision Conference (BMVC), pp. 1–11
18. Callut J, Dupont P, Saerens M (2011) Sequence classification in the Jensen-Shannon embedding. In: International conference on machine learning
19. Scholkopf B, Smola AJ (2001) Learning with kernels-support vector machines, regularization, optimization and beyond. MIT Press, Cambridge

Rough Set Rule-Based Technique for the Retrieval of Missing Data in Malaria Diseases Diagnosis

B.S. Panda, S.S. Gantayat and Ashok Misra

Abstract Malaria disease is a major tropical public health problem in the world. The diagnosis of this type of tropical diseases involves several levels of uncertainty and imprecision. It causes severe infection to the brain and prevents brain from its proper functioning. Hence prior detection of the malaria is much essential. Soft Computing Techniques provide excellent methodologies to process the medical data and help medical experts in finding out the nature of illness and to take decision. True data set collection, feature squeezing, and classification are the basic steps followed in designing an expert system. The designed expert system acts with intelligence, prevents erroneous decisions, and produces sharp results in time. This paper discusses on malaria investigation with missing data using rough set rule-based soft computing technique.

Keywords Accuracy · Malaria · Missing data · Medical diagnosis · Rough set · Rule set

1 Introduction

In India, malaria constitutes a great threat to the health of many communities. The harmful effects of malaria parasites to the human body cannot be under estimated. Malaria is a parasitic disease caused mainly by species of Anopheles mosquitoes. The

B.S. Panda (✉)
MITS, Rayagada, Odisha, India
e-mail: bhavanipanda@yahoo.com

S.S. Gantayat
GMRIT, Rajam, Andhra Pradesh, India
e-mail: sasankosekhar.g@gmrit.org

A. Misra
CUTM, Paralakhemundi, Odisha, India
e-mail: amisra1972@gmail.com

© The Author(s) 2015   59
N.B. Muppalaneni and V.K. Gunjan (eds.), *Computational Intelligence
in Medical Informatics*, Forensic and Medical Bioinformatics,
DOI 10.1007/978-981-287-260-9_6

emergence of information technology (IT) has opened unprecedented opportunities in health care delivery system as the demand for intelligent and knowledge-based systems has increased as modern medical practices become more knowledge-intensive. The diagnosis of tropical diseases involves several levels of uncertainty and imprecision [1].

The task of disease diagnosis and management is complex because of the numerous variables involved in data sets, which may contain a lot of imprecision and uncertainties. Patients cannot describe exactly how they feel, doctors and nurses cannot tell exactly what they observe, and laboratories results are dotted with some errors caused either by the carelessness of technicians or by malfunctioning of the instrument. Medical researchers cannot precisely characterize how diseases alter the normal functioning of the body [2, 3]. All these complexities in medical practice make traditional quantitative approaches of analysis inappropriate. Computer tools help to organize, store, and retrieve missing appropriate medical knowledge needed by the practitioner in dealing with each difficult case and suggesting appropriate diagnosis and decision-making technique.

In this article, we provided a rule-based soft computing technique for the diagnosis of malaria to find the appropriate (missing) symptoms to diagnosis the malaria. This technique is based on clinical observations, medical diagnosis, and the expert's knowledge. The objective of the system is to provide a decision support platform to medical researchers, physicians, and other healthcare practitioners in malaria endemic regions. In addition, the system will assist medical personnel in the tedious and complication task of diagnosing and further provide a scheme that will assist medical personnel especially in rural areas, where there are shortages of doctors, thereby, and offering primary health care to the people.

2 Literature Review

Missing data are questions without answers or variables without observation. Even a small percentage of missing data can cause serious problems with the analysis leading to draw wrong conclusions and imperfect knowledge. There are many techniques developed in the literature to manipulate the knowledge with uncertainty and manage data with incomplete items, but no results, and the results are not of the same type and absolutely better than the others [4–6].

To handle such problems, researchers are trying to solve it in different approaches and then proposed to handle the information system in their way. We know that the attribute values are important for information processing in a data set or information table. In the field of databases, various efforts have been made for the improvement and enhance of database or information table query process to retrieve the data. The methodology followed by different approaches such as fuzzy sets [7, 8], rough sets [9, 10], Boolean logic, possibility theory, statistically similarity [11], etc.

3 Rough Sets

3.1 Definition and Notations

The concept of rough set is another approach to deal with imperfect knowledge. It was introduced by Pawlak in 1982 [10]. From a philosophical point of view, rough set theory is a new approach to deal vagueness and uncertainty, and from a practical point of view, it is a new method of data analysis.

This method has the following important advantages:

- It provides efficient algorithms for finding hidden patterns in data;
- It finds reduced set of data (data reduction);
- It evaluates significance of data;
- It generates minimal set of decision rules from data;
- It is easy to understand;
- It offers straightforward interpretation of results;
- It can be used in both qualitative and quantitative data analyses; and
- It identifies relationship that would not be found by using statistical methods.

Let $R \subseteq UxU$ denote an equivalence relation on U, that is, R is a reflexive, symmetric, and transitive relation. The equivalence class of an element $x \in U$ with respect to R is the set of elements $y \in U$ such that xRy. If two elements x, y in U belong to the same equivalence class then we say that x and y are indistinguishable with respect to relation R.

Given an arbitrary set $A \subseteq U$, it may not be possible to describe "A" precisely in the approximation space apr$(R) = (U, R)$ instead one may only characterize "A" by a pair of lower and upper approximations. This leads to the concept of rough sets (Fig. 1).

Fig. 1 Rough set approach

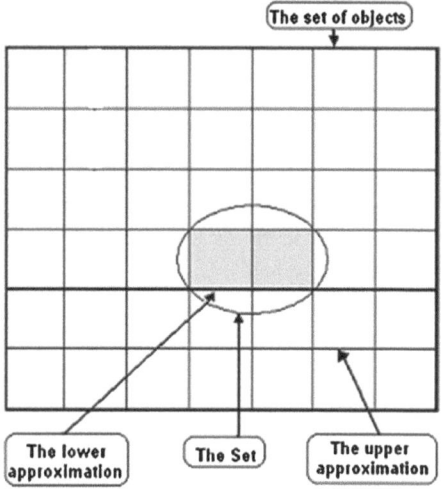

We define,

$$\underline{R}A = \cup \{Y \in U/R \, : \, Y \subseteq A\};$$

and

$$\overline{R}A = \cup \{Y \in U/R \, : \, Y \cap A \neq \varphi\}.$$

$\underline{R}A$ and $\overline{R}A$ are respectively called the R-lower and R-upper approximation of A with respect to R.

It can be noted that

$$\underline{R}A = \{x \in U : [x_R] \subseteq A\}$$

and

$$\overline{R}A = \{x \in U : [x]_R \cap X \neq \varphi\}.$$

The set $BN_R(A) = \overline{R}A - \underline{R}A$ is called the R-*boundary* of A. The set $\underline{R}A$ consists of all those elements of U which can with certainty be classified as elements of A, employing the knowledge R. The set $\overline{R}A$ consists of all those elements of U which can possibly be classified as elements of A, employing the knowledge R. Set $BN_R(A)$ is the set of elements which cannot be classified as either belonging to A or belonging to $-A$ having the knowledge R. We say that a set A is R-*definable* if and only if $\underline{R}A = \overline{R}A$. Otherwise A is said to be R-*rough*.

The borderline region is the undecidable area of the universe. We say that X is *rough* with respect to R if and only if $\underline{R}X \neq \overline{R}X$, equivalently $BN_R(X) \neq \phi$. X is said to be R-*definable* if and only if $\underline{R}X = \overline{R}X$, or $BN_R(X) = \phi$.

3.2 Application of Rough Set

We briefly highlight few applications of Rough set theory [12]. We again feel discussing on various applications will increase the length of paper.

- Representation of uncertain or imprecise knowledge.
- Empirical learning and knowledge acquisition from experience.
- Knowledge analysis.
- Analysis of conflicts.
- Evaluation of the quality of the available information with respect to its consistency and the presence or absence of repetitive data patterns.
- Identification and evaluation of data dependencies.
- Approximate pattern classification.
- Reasoning with uncertainty.
- Information-preserving data reduction.

4 Malaria Data Set with Missing Attribute

Table 1 shows the attributes *Temperature, Headache, Nausea, Vomiting, Joint Pain, and Body Weakness* and with the decision *Malaria*. However, many real-life data sets are incomplete. The missing attribute value is denoted by "?".

From the above, the different classes can be generated using rough set concept as follows.

Temperature = {C01, C06, C07, C10, C13, C15}, {C02, C05, C08, C09, C12}, {C03, C14}, {C04}, **{C11}}**

Headache = {{C01, C02, C04, C05, C07, C09, C10, C15}, {C03, C06, C13}, **{C08}**, {C11, C12, C14}}

Nausea = {{C01, C02, C03, C05, C15}, **{C04, C08}**, {C06, C07, C09, C10, C12, C13, C14}, {C11}}

Vomiting = {{C01, C03, C04, C06}, **{C02}**, {C05, C07, C08, C09, C10, C13, C15}, {C11, C12, C14}}

Joint_Pain = {{C01, C03, C15}, {C02, C05, C08, C09, 12}, {C04, C07, C11, C14}, **{C06, C10, C13}}**

Body_Weakness = {{**C01, C05**}, {C02, C08, C09, C13}, {C03, C04, C07, C12, C14, C15}, {C06, C10}, **{C11}}**

Malaria = {{C01, C05, C06, C10, C13, C15}, {C02, C03, C04, C07, C08, C09, C11, C12, C14}}

In the above classification, the bold classes are the missing data, which will be filled in the currently proposed technique.

Here, the rules are in the following two classes for Malaria types "Yes" or "No".

If the symptoms lie in any of the temperature, head ache, nausea, vomiting, joint pain, body weakness, then the Malaria type is either "Yes" or "No".

5 Proposed System

The proposed system for malaria diagnosis uses rough set rule-based system. The steps included in the proposed system are given below in Fig. 2.

Algorithm:

Step 1: **Malaria Data Set** Collect the Malaria data set which includes relevant objects and attributes.

Setp 2: **Divide Data Set** Divide the malaria set into training and testing sets using split factor of 60 and 40 %.

Step 3: **Cut Set** Form cut set for the training set which represents the nonoverlapping subsets thus divides the attributes into set of intervals. Cuts act as a boundary values representing those intervals.

Table 1 Malaria data set with missing attributes

Case#	Temperature	Headache	Nausea	Vomiting	Joint_Pain	Body_Weakness	Malaria
C01	Mild	Mild	Mild	Mild	Mild	?	No
C02	Moderate	Mild	Mild	?	Moderate	Moderate	Yes
C03	Severe	Moderate	Mild	Mild	Mild	Severe	Yes
C04	Very severe	Mild	?	Mild	Severe	Severe	Yes
C05	Moderate	Mild	Mild	Moderate	Moderate	?	No
C06	Mild	Moderate	Moderate	Mild	?	Mild	No
C07	Mild	Mild	Moderate	Moderate	Severe	Severe	Yes
C08	Moderate	?	?	Moderate	Moderate	Moderate	Yes
C09	Moderate	Mild	Moderate	Moderate	Moderate	Moderate	Yes
C10	Mild	Mild	Moderate	Moderate	?	Mild	No
C11	?	Severe	Severe	Severe	Severe	Very severe	Yes
C12	Moderate	Severe	Moderate	Severe	Moderate	Severe	Yes
C13	Mild	moderate	Moderate	Moderate	?	Moderate	No
C14	Severe	Severe	Moderate	Severe	Severe	Severe	Yes
C15	Mild	Mild	Mild	Moderate	Mild	Severe	No

Fig. 2 Flow chart of the rule-based system

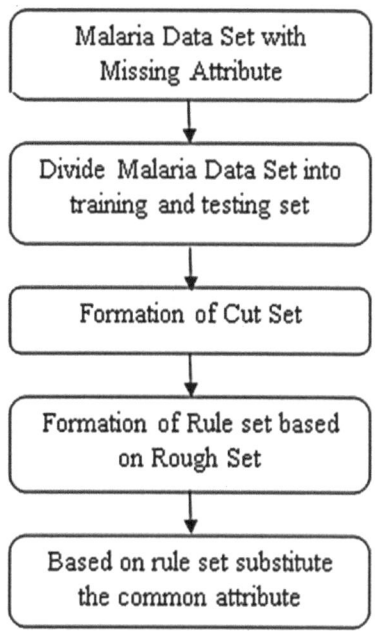

Step 4: **Rule Set** Rules are formed from the attribute by common attribute values through logical AND. It also states the result class to which a particular rule point to.

Step 5: **Substitute Common Attribute** One of the simplest methods to handle missing attribute values, such values are replaced by the most common value of the attribute. In different words, a missing attribute value is replaced by the most probable known attribute value, where such probabilities are represented by relative frequencies of corresponding attribute values.

Note: The rules generated here can be extended with the logical operators OR and NOT for a large dataset.

6 Cut Set for Overlapping Data

Cut set defines the mechanism of decomposing attribute value set in a dataset or information table. The attributes are discretized to generate a set of intervals. These are edge points to highlight the intervals. In case of representative attributes, cuts stand for the nonoverlapping subsets of actual values. Table 2 represents cut set consists of attributes for which the cuts are created. Size represents the quantity of cuts created. In the following table, the missing attributes are represented by "?".

Table 2 After applying cut set

Case#	Temperature	Headache	Nausea	Vomiting	Joint_Pain	Body_Weakness	Malaria
C01	Mild	Mild	Mild	Mild	Mild	?	*No*
C02	Moderate	Mild	Mild	?	Moderate	Moderate	*Yes*
C04	Very severe	Mild	?	Mild	Severe	Severe	*Yes*
C05	Moderate	Mild	Mild	Moderate	Moderate	?	*No*
C06	Mild	Moderate	Moderate	Mild	?	Mild	*No*
C08	Moderate	?	?	Moderate	Moderate	Moderate	*Yes*
C10	Mild	Mild	Moderate	Moderate	?	Mild	*No*
C11	?	Severe	Severe	Severe	Severe	Very severe	*Yes*
C13	Mild	Moderate	Moderate	Moderate	?	Moderate	*No*

Cut set states the result attribute in terms of other attributes, which is a feature reduction based on Boolean logics of algebra.

7 Rough Set Rule-Based Technique

The rough set rule-based technique usually have several antecedents that are combined using logical operators, such as AND, OR, and NOT, though their definitions tend to vary. In the rules, AND simply uses minimum weight of all the antecedents, while OR uses the maximum value, the IF part is called the "antecedent," and the THEN part is called the "consequent" [13, 14]. The conditions are the set of attributes such as temperature, head ache, nausea, vomiting, joint pain, body weakness, etc.

There are different approaches to generate rules, direct and indirect methods. Direct methods generate rule from training data like sequential covering algorithms. Indirect methods build the classification model from which rule are extracted, e.g., decision tree, neural network, genetic algorithms, etc.

The rule can be assessed by its coverage and accuracy. *Coverage*, if all the antecedents in a rule R hold true for a given D. *Accuracy*, is the number of tables that are correctly classified in the coverage. The coverage and accuracy are determined by the following formula.

$$\text{Coverage}(R) = \frac{n_{\text{covers}}}{|D|} \qquad \text{Accuracy}(R) = \frac{n_{\text{correct}}}{n_{\text{covers}}}$$

Information gain is a supervised feature selection method. It is a method that measures the decrease in entropy when the feature is given rather than not given. It can generalize to any number of classes. The entropy of a discrete random variable X with a probability distribution $p(x)$ is defined as

$$H(p) = \sum_{x \in X} p(x) \log p(x)$$

These formulas are determined by aid of both the expert doctors in the field of tropical medicine and literature. Some of the rules (Rules 1, Rules 2, Rules 5, Rule 8, ...Rule 15) can be interpreted as follows:

Rule 1:   IF *temperature* = mild and *headache* = mild and *nausea* = mild and *vomiting* = mild and *joint pain* = mild and *body weakness* = ? THEN *malaria* = No.

Rule 2:   IF *temperature* = moderate and *headache* = mild and *nausea* = mild and *vomiting* = ? and *joint pain* = moderate and *body weakness* = moderate THEN *malaria* = Yes.

Rule 3:   IF *temperature* = severe and *headache* = moderate and *nausea* = mild and *vomiting* = mild and *joint pain* = mild and *body weakness* = severe THEN *malaria* = Yes.

Rule 5:   IF *temperature* = moderate and *headache* = mild and *nausea* = mild and *vomiting* = moderate and *joint pain* = moderate and *body weakness* = ? THEN *malaria* = No.

Rule 8:   IF *temperature* = moderate and *headache* = ? and *nausea* = ? and *vomiting* = moderate and *joint pain* = moderate and *body weakness* = moderate THEN *malaria* = Yes.

Rule 15:   IF *temperature* = mild and headache = mild and *nausea* = mild and *vomiting* = moderate and *joint pain* = mild and *body weakness* = severe THEN *malaria* = Yes.

The notion of rough set [15–17] is used to classify the patients with malaria or not and also the other attributes which gives a group of similar information in the data set. Here the other rules are not mentioned to avoid the complexity and confusion with the given rules.

8 Results and Discussion

To implement the concept of rough set rule-based technique to simplifying the diagnosis of malaria and impute the missing attributes, we have considered 15 data sets with six attributes out of 20 real data sets with ten attributes collected from different doctors. The technique used on knowledge of domain experts (five medical doctors), applied with rough set theory. Rough set classification is utilized to remove uncertainty, ambiguity, and vagueness inherent in medical diagnosis.

The bold attributes in the above table show the unmatched attribute with the actual attributes, to measure the accuracy using the formula we found 70 % of accuracy, i.e., out of 10 cases 7 cases are matched. After imputation of common attribute using rough set rule-based technique, the following tables 3 and 4 shows the results.

Table 3 Imputation result for missing data with actual data

Case#	Common attribute	Actual attribute	Diagnosis
C01	Mild	Mild	No
C02	**Moderate**	Mild	Yes
C04	**Severe**	Mild	Yes
C05	Moderate	Moderate	No
C06	Mild	Mild	No
C08	Moderate	Moderate	Yes
C10	Mild	Mild	No
C11	Severe	Severe	Yes
C13	**Moderate**	Mild	No

Table 4 The missing data filled with observed data

Case#	Temperature	Headache	Nausea	Vomiting	Joint_Pain	Body_Weakness	Diagnosis
C01	Mild	Mild	Mild	Mild	Mild	**Mild**	No
C02	Moderate	Mild	Mild	*Moderate*	Moderate	Moderate	Yes
C03	Severe	Moderate	Mild	Mild	Mild	Severe	Yes
C04	Very severe	Mild	*Severe*	Mild	Severe	Severe	Yes
C05	Moderate	Mild	Mild	Moderate	Moderate	**Moderate**	No
C06	Mild	Moderate	Moderate	Mild	**Mild**	Mild	No
C07	Mild	Mild	Moderate	Moderate	Severe	Severe	Yes
C08	Moderate	**Moderate**	**Moderate**	Moderate	Moderate	Moderate	Yes
C09	Moderate	Mild	Moderate	Moderate	Moderate	Moderate	Yes
C10	Mild	Mild	Moderate	Moderate	**Mild**	Mild	No
C11	**Severe**	Severe	Severe	Severe	Severe	Very severe	Yes
C12	Moderate	Severe	Moderate	Severe	Moderate	Severe	Yes
C13	Mild	moderate	Moderate	Moderate	*Moderate*	Moderate	No
C14	Severe	Severe	Moderate	Severe	Severe	Severe	Yes
C15	Mild	Mild	Mild	Moderate	Mild	Severe	No

The bold word in the above table shows the replacement of approximated information with the missing data in Table 4.

After replacing proper values to the missing data, we get the different classes from the above table as follows.

Temperature = {C01, C06, C07, C10, C13, C15}, {C02, C05, C08, C09, C12}, {C03, **C11**, C14}, {C04}}

Headache = {{C01, C02, C04, C05, C07, C09, C10, C15}, {C03, C06, **C08**, C13}, {C11, C12, C14}}

Nausea = {{C01, C02, C03, C05, C15}, {C06, C07, **C08**, C09, C10, C12, C13, C14}, {**C04**, C11}}

Vomiting = {{C01, C03, C04, C06}, {**C02**, C05, C07, C08, C09, C10, C13, C15}, {C11, C12, C14}}

Joint_Pain = {{C01, C03, **C06**, **C10**, C15}, {C02, C05, C08, C09, C12, **C13**}, {C04, C07, C11, C14}}

Body_Weakness = {{C02, **C05**, C08, C09, C13}, {C03, C04, C07, C12, C14, C15}, {**C01**, C06, C10}, {C11}}

Malaria = {{C01, C05, C06, C10, C13, C15}, {C02, C03, C04, C07, C08, C09, C11, C12, C14}}

After the imputation, the different classes for individual attributes are reassigned with the currently added new data.

From the above table, it is clear that the classes generated for the decision making are reducing after imputing the missing values by using the rule generation method (Table 5).

The analysis of the classification shows the following result.

Table 5 Comparison of before and after imputing missing values

Symptom types	No. of classes before imputing missing values (A)	No. of classes after imputing missing values (B)	No. of changes in classes |(A-B)|
Temperature	5	4	1
Headache	4	3	1
Nausea	4	3	1
Vomiting	4	3	1
Joint_Pain	4	3	1
Body_Weakness	5	4	1

9 Conclusion

Malaria investigation using rough set rule-based soft computing technique shows good imputation and coverage. It has handled 15 cases and 6 attributes. The above method handles attributes even in the presence of absent values. In future, the same method can be used to investigate other diseases by replacing the absent values with common attribute value. This approach can be combined with other classifiers to enhance further to retrieve the missing data in a data set or information table. This technique can be enhanced by using rough logic, which is a future direction.

References

1. Uzoka FME, Osuji J, Obot O (2010) Clinical decision support system (DSS) in the diagnosis of malaria: a case comparison of two soft computing methodologies. Expert Syst Appl 38:1537–1553
2. Szolovits P, Patil RS, Schwartz WB (1988) Artificial intelligent in medical diagnosis. J Intern Med 108:80–87
3. Szolovits P (1995) Uncertainty and decision in medical informatics. Methods Inf Med 34:111–121
4. Little RJ, Rubin DB (2002) Statistical analysis with missing data, 2nd edn. Wiley, New York
5. Kantadzic M (2003) Data mining: concepts, models, methods and algorithms. Wiley, New York
6. Gantayat SS, Misra A, Panda BS (2013) A study of incomplete data—a review. In: LNCS Springer FICTA-2013, pp 401−408. ISBN: 978-3-319-02930-6
7. Zadeh LA (1965) Fuzzy sets. Inf Control 8:338–353
8. Zadeh LA (1973) Outline of a new approach to the analysis of complex system and decision processes. IEEE Trans Syst Man Cybern 3:28–44
9. Grzymala-Busse J (1988) LERS-a system for learning from examples based on rough sets. J Intell Rob Syst 1:3–16
10. Pawlak Z (1982) Rough sets. J Inf Comp Sci II:341–356
11. Devlin H, Devlin JK (2007) Decision support system in patient diagnosis and treatment. Future Rheumatol 2:261–263
12. Panda BS, Abhishek R, Gantayat SS (2012) Uncertainty classification of expert systems—a rough set approach. In: ISCON proceedings with IJCA. ISBN: 973-93-80867-87-0
13. Grzymala-Busse J (1988) Knowledge acquisition under uncertainty—a rough set approach. J Intell Rob Syst 1:3–16
14. Panda BS, Gantayat SS, Misra A (2013) Rough set approach to development of a knowledge-based expert system. Int J Adv Res Sci Technol (IJARST) 2(2):74–78. ISSN: 2319-1783
15. Pawlak Z (1991) Rough sets-theoretical aspects of reasoning about data. Kluwer Academic Publishing, Boston
16. Pawlak Z, Skowron A (2007) Rough sets- some extensions. Inf Sci 177(1):28–40
17. Pawlak Z (1996) Why rough sets, fuzzy systems. In: Proceedings of the fifth ieee international conference, vol 2

Automatic Image Segmentation for Video Capsule Endoscopy

V.B. Surya Prasath and Radhakrishnan Delhibabu

Abstract Video capsule endoscopy (VCE) has proven to be a pain-free imaging technique of gastrointestinal (GI) tract and provides continuous stream of color imagery. Due to the amount of images captured automatic computer-aided diagnostic (CAD) methods are required to reduce the burden of gastroenterologists. In this work, we propose a fast and efficient method for obtaining segmentations of VCE images automatically without manual supervision. We utilize an efficient active contour without edges model which accounts for topological changes of the mucosal surface when the capsule moves through the GT tract. Comparison with related image segmentation methods indicate we obtain better results in terms of agreement with expert ground-truth boundary markings.

Keywords Capsule endoscopy · Image segmentation · Active contours · CAD

1 Introduction

Oesophagel cancer rate in India is one of the highest in the world. Although low and stable incidence and mortality rates from colorectal cancers were observed in India [1], these rates were associated with a low 5-year relative survival rate [2]. This low survival rate suggests severe deficiencies in early diagnosis and effective treatment in India. Since as noted in [2] population-based screening of colorectal

V.B. Surya Prasath (✉)
University of Missouri-Columbia, Columbia, MO 65211, USA
e-mail: prasaths@missouri.edu

R. Delhibabu
Cognitive Modeling Lab, IT University Innopolis, Kazan, Russia
e-mail: delhibabur@ssn.edu.in

R. Delhibabu
Department of CSE, SSN Engineering College, Chennai, India

R. Delhibabu
Machine cognition lab, Kazan Federal University, Kazan, Russia

© The Author(s) 2015
N.B. Muppalaneni and V.K. Gunjan (eds.), *Computational Intelligence in Medical Informatics*, Forensic and Medical Bioinformatics,
DOI 10.1007/978-981-287-260-9_7

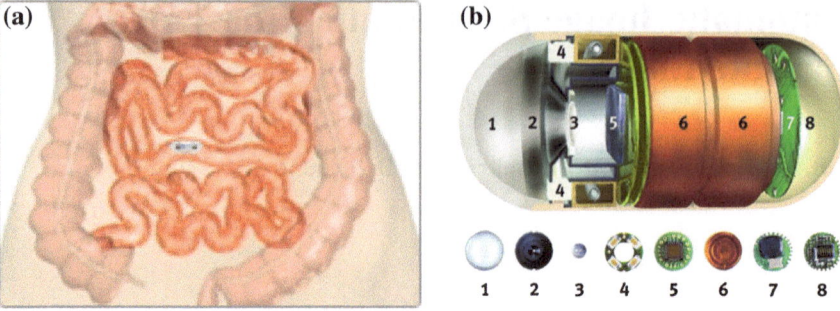

Fig. 1 Wireless capsule endoscopy: **a** The patient swallows the capsule and capsule travels through the tubular intestinal path. The imaging is done through a *circular lens* and at periodic intervals. **b** Pillcam Capsule parts (used with permission from). *1* Optical dome. *2* Lens holder. *3* Lens. *4* Illuminating LEDs. *5* CMOS imager. *6* Battery. *7* ASIC transmitter. *8* Antenna

cancer is not cost-effective, given the low burden of colorectal cancer, early diagnosis and adequate treatment using imaging techniques are important.

VCE introduced at the turn of this millennium [3] paved the way for painless non-invasive way to image the gastrointestinal tract, see Fig. 1. A wireless capsule consist of a tiny imaging device provides continuous video stream of the inner mucosa-lumen tubular structure. Typically, these VCE exam for each patient consist of 8 h of video which is around 55,000 frames. This puts an enormous burden on the gastroenterologists since examining and reviewing the video from capsule endoscopy requires concentration for a long durations. Thus, automatic CAD methods are required which can help in analysis and diagnosis [4–6], see [7] for a recent review. Computerized video analysis algorithms can reduce the time required to review VCE exams and can augment the decision making processes.

Lumen detection in other imaging modalities have been studied by some researchers. In [8, 9] colonoscopy images are considered and adaptive thresholding type techniques were utilized. The wireless capsule optical system usually consist of flashing LEDs, as opposed to fixed lighting in colonoscopy, and an imaging sensor which captures the images at predefined time intervals. Thus, thresholding type methods may not capture strong intensity variations between frames. In this work, we study an automatic image segmentation method for VCE imagery to segment lumen boundaries without any manual supervision. For this purpose we start with the well-known image segmentation method based on active contours without edges [10]. This method involves level sets which are known to handle topological changes and thus is suitable for mucosa deformations which occur in VCE data [11]. Though efficient, the traditional implementation was not fast for real-time imaging and is restricted to takes 0.32 s/frame (of size 512 × 512). To speed up the active contour segmentation further, in this paper we propose a fast and efficient implementation of the active contour model which obtains good segmentations. Compared with some of the classical and contemporary image segmentation techniques we obtain better segmentations when benchmarked with marked ground-truth boundaries.

The rest of the paper is organized as follows. Section 2 gives the proposed fast active contour based segmentation method. Section 3 provides experimental results on various VCE images supporting our proposed approach. Finally, Sect. 4 concludes the paper.

2 Active Contours for Lumen-Mucosa Separation

2.1 Active Contour Without Edges

Chan and Vese [10] studied the active contour without edges (ACWE) method which gives better segmentation results than other active contour based schemes. ACWE is a modified version of the Mumford-Shah functional [12] and is an alternating minimization scheme. Implementation of the nonlinear minimization scheme is undertaken using the level sets methodology [13]. Suppose that the input image $u : \Omega \subset \mathbb{R}^2 \to \mathbb{R}$ where Ω represents the image domain, usually a rectangle. Then the piecewise constant ACWE is given by

$$\min_{\phi, c_1, c_2} \varepsilon(\phi, c_1, c_2) = \mu \int_\Omega \delta(\phi)|\nabla\phi|dx + \lambda \int_\Omega |u - c_1|^2 H(\phi)dx$$
$$+ \lambda \int_\Omega |u - c_2|^2 (1 - H(\phi))dx \tag{1}$$

where $\lambda > 0$ and $\mu \geq 0$ are given fixed parameters, c_1, c_2 represent regions inside and outside the level set function ϕ respectively. The function $H(z) := 1$ if $z \geq 0$, $H(z) := 0$ if $z < 0$ is the Heaviside function, and $\delta(z) := \frac{d}{dz}H(z)$ is the Dirac delta function in the sense distributions. In [10], the Euler-Lagrange equation of the functional (1), which is a necessary condition for a minimizer triplet (ϕ, c_1, c_2) to satisfy, is implemented. That is, the following nonlinear PDE is solved for ϕ and c_1, c_2:

$$\frac{\partial\phi}{\partial t} = \delta_\varepsilon(\phi)\left[\mu \, div\left(\frac{\nabla\phi}{|\nabla\phi|}\right) - \lambda_1(u - c_1) + \lambda_2(u - c_2)\right] \tag{2}$$

with

$$c_1 = \frac{\int_\Omega uH_\varepsilon(\phi)dx}{\int_\Omega uH_\varepsilon(\phi)dx}, \quad c_2 = \frac{\int_\Omega u(1 - H_\varepsilon(\phi))dx}{\int_\Omega u(1 - H_\varepsilon(\phi))dx} \tag{3}$$

Here, δ_ε, H_ε represent the regularized versions of the dirac delta and Heaviside functions respectively.

2.2 Fast Implementation

The discretized version of the ACWE scheme (2) can be obtained using finite differences and the method proposed in [10] follows a standard explicit central differences for the spatial variable and forward difference for time variable [11]. In this work, we modify the standard scheme by using a non-standard discretization and our method is written as follows: Let $\phi_{i,j}^{k} = \phi(n\Delta t, x_i, y_j)$ be an approximation of $\phi(t, x, y), k \geq 0, \phi^0 = \phi_0$ and Δt is the time step size. Consider the first order standard differences

$$\Delta_-^x \phi_{i,j} = \phi_{i,j} - \phi_{i-1,j}, \quad \Delta_+^x \phi_{i,j} = \phi_{i+1,j} - \phi_{i,j}$$
$$\Delta_-^y \phi_{i,j} = \phi_{i,j} - \phi_{i,j-1}, \quad \Delta_+^y \phi_{i,j} = \phi_{i,j+1} - \phi_{i,j}.$$

Then the algorithm for solving the ACWE PDE (2) is an alternating iterative scheme:

1. Fix μ, λ, ε knowing ϕ^k, compute $c_1(\phi^k)$, $c_2(\phi^k)$ from (3).
2. Compute ϕ^{k+1} by the following discretization and linearization of (2) in ϕ:

$$
\begin{aligned}
\frac{\phi_{i,j}^{k+1} - \phi_{i,j}^{k}}{\Delta t} = \delta_h\left(\phi_{i,j}^{k}\right) &\left[\frac{\mu}{h^2}\Delta_-^x\left(\frac{\Delta_+^x \phi_{i,j}^{k+1}}{\sqrt{\left(\Delta_+^x\phi_{i,j}^k\right)^2/h^2 + \left(\phi_{i,j+1}^k - \phi_{i,j-1}^k\right)^2/(2h)^2}} \right) \right. \\
&+ \frac{\mu}{h^2}\Delta_-^y\left(\frac{\Delta_+^y \phi_{i,j}}{\sqrt{\left(\phi_{i+1,j}^k - \phi_{i-1,j}^k\right)^2/(2h)^2 + \left(\Delta_+^y\phi_{i,j}^k\right)^2/h^2}} \right) \\
&\left. - \lambda_1\left(u_{i,j} - c_1\left(\phi^k\right)\right)^2 + \lambda_2\left(u_{i,j} - c_2\left(\phi^k\right)\right)^2 \right]
\end{aligned}
$$

$$(4)$$

The modified finite difference scheme (4) involve only first order discrete derivatives and theoretical convergence of the above discretization will be reported elsewhere.

3 Experimental Results

3.1 Setup and Parameters

The core operations of the schemes are implemented in C with *mex* interface and MATLAB® R2012a is used for visualization purposes. Our fast scheme takes about 0.1 s (40 iterations) for a 512×512 image compared to 0.32 s in a previous work [11]. This means on average the segmentation can be run for 10 frames per second (fps), making it very attractive for endoscopic image analysis and diagnosis. Enabling further code optimization via GPU may significantly speed up the main loop of the program. Moreover, instead of having a fixed number of iterations for segmentation, we devised a tolerance check to stop the scheme at convergence. For example, $\left|\phi^{k+1} - \phi^k\right| < 10^{-6}$ was set, and we observed convergence of the iterative scheme (4) in less than 40 iterations for most of the images. Thus, our scheme takes average runtime of 0.1 s/frame using non-optimized MATLAB code on 2.3 GHz Intel Core i7, 8 GB RAM laptop to obtain a segmentation of lumen and mucosa.

The ACWE parameters $\lambda_1 = \lambda_2 = 1$, and $\varepsilon = 10^{-6}$ are fixed for all the experiments reported here. This gives equal weight to both inside and outside regions of the zero level set. The μ parameter in Eq. (4) controls how regular the final active contour is, and hence, affects the final segmentation in the experiments. After conducting extensive experiments with respect to this parameter, we fixed it at $\mu = 0.2$, and this seems to work well for most of the images in VCE data. The time step parameter $\Delta t = 0.5$, step size for the finite difference grid $h = 1$ are fixed throughout all the experiments.

3.2 Comparison Results

The first example in Fig. 2a shows the segmentation based on the proposed scheme. The original image is a typical normal mucosa of the human colon captured in vivo by the wireless capsule endoscopy device. To obtain good segmentation we always start with the initialization shown in Fig. 2a, this type of fine mesh kind of

(a)	(b)	(c)	(d)	(e)

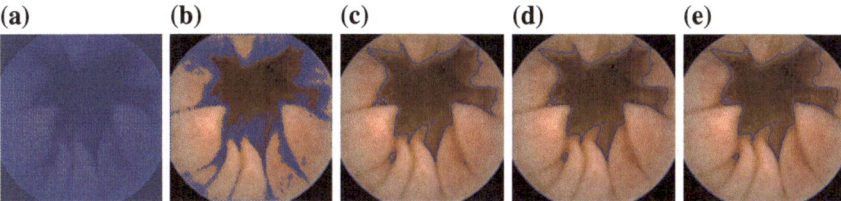

Fig. 2 Endoscopic image mucosa segmentation. **a** Input image and initialization of the modified ACWE scheme, see Sect. 2.2. **b–d** Intermediate segmentation results at time $t = 20$, 40 and 60, curve laid on *top* of the original image for visualization. **e** Final segmentation at $T = 80$. The segmentation algorithm usually converges within $t = 40$

initialization provided better segmentations. Moreover, we observed experimentally that, with a single or seed of coarse circles based initializations the proposed scheme took more iterations to converge. Figure 2b–d shows the intermediate active contours of the segmentation scheme. The final segmentation result in Fig. 2e shows the mucosal folds and the existence of a clear boundary between the lumen section of the image.

Related schemes from literature are also compared in terms segmentation accuracy in mucosa-lumen differentiation. Figure 3 we show a comparison result of other segmentation schemes from the literature against our scheme. The following methods are compared with the proposed scheme (Table 1).

As can be seen by comparing, white curves superimposed, in Fig. 3g of the proposed approach's segmentation result, we see that it captures mucosa regions efficiently even if the image is of low contrast. Texture based active contour (TAC) fails completely and mean-shift obtains spurious lumen regions. On the other hand,

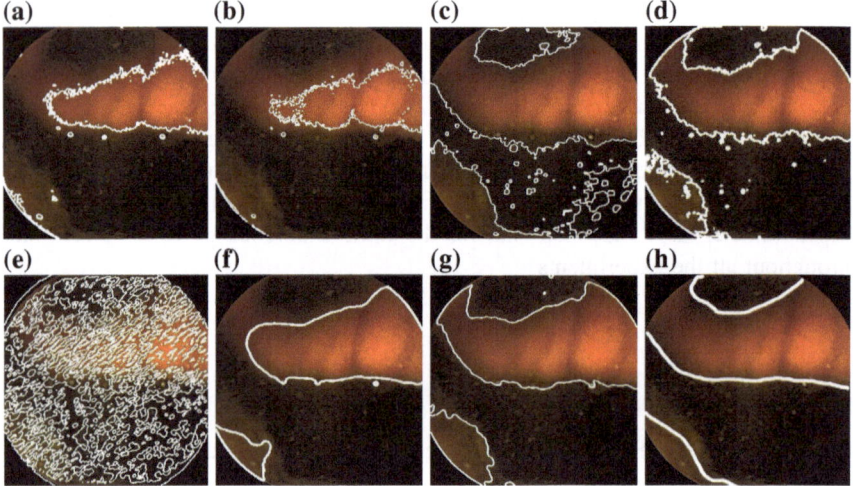

Fig. 3 Comparison with other segmentation schemes: **a** Otsu thresholding based scheme [8]. **b** Adaptive progressive thresholding scheme [9]. **c** Mean shift [17]. **d** ACWE [10]. **e** Texture-ACWE [18]. **f** Global minimization model [19] with Chan-Vese energy. **g** Our fast active contour. **h** Expert (gastroenterologist) segmentation. Dice values (κ) is given for each result and our fast implementation obtained better agreement with ground-truth (GT) boundaries

Table 1 .

(a) Otsu thresholding based scheme (OT)	[8]	Fig. 3a
(b) Adaptive progressive thresholding (APT)	[9]	Fig. 3b
(c) Mean shift (MS)	[17]	Fig. 3c
(d) Chan-Vese scheme (ACWE)	[10]	Fig. 3d
(e) Texture based active contour (TAC)	[18]	Fig. 3e
(f) Fast global minimization (FGM)	[19]	Fig. 3f

Table 2 Dice similarity coefficient (κ, Eq. (5)) values for automatic segmentations when compared with manual ground truth for different schemes in Fig. 3

Sub-figure	(a)	(b)	(c)	(d)	(e)	(f)	Our
Ref.	[8]	[9]	[17]	[10]	[18]	[19]	Result
	0.6592	0.4987	0.8202	0.8182	0.3017	0.7444	**0.8793**

results of other schemes such as APT, ACWE give inaccurate segmentations of the mucosa folds. We use the Dice Similarity Coefficient,

$$\kappa(A, B) = \frac{2|A \cap B|}{|A| + |B|}, \tag{5}$$

where A and B are automatic and manual segmentations respectively, and $|X|$ denotes the total number of pixels in a set X. Dice values are in [0,1.0] with values closer to one indicate good overlap between manual and automatic segmentations. Table 2 provides Dice values for different schemes corresponding to Fig. 3. As can be seen our method obtains the highest Dice value and confirms visual comparison seen with the GT image. Similar active contour methods utilized in other imaging modalities [14–16] can also be adapted for VCE imagery and defines our future work in this direction.

4 Conclusion

In this paper, we consider automatic segmentation of video capsule endoscopy imagery. By utilizing an active contour without edges with a fast implementation we obtain meaningful segmentations of the gastrointestinal tract imaged by the capsule endoscopy. Numerical implementation of the proposed scheme is carried out with finite differences and provides efficient segmentation results. Compared with other related image segmentation methods our fast implementation obtains better results when compared to ground-truth marked by gastroenterologists. We believe automatic computer aided diagnostic methods can provide relief to the big-data handling associated with reading video capsule endoscopy imagery and further methods for augmenting the diagnostic capabilities are required currently.

References

1. Mohandas KM (2011) Colorectal cancer in india: controversies, enigmas and primary prevention. Indian J Gastroenterol 30(1):3–6
2. Pathy S, Lambert R, Sauvaget C, Sankaranarayanan R (2012) The incidence and survival rates of colorectal cancer in India remain low compared with rising rates in east asia. Dis Colon Rectum 55(8):900–906

3. Iddan G, Meron G, Glukhovsky A, Swain F (2000) Wireless capsule endoscopy. Nature 405 (6785):417
4. Figueiredo PN, Figueiredo IN, Prasath S, Tsai R (2011) Automatic polyp detection in pillcam colon 2 capsule images and videos: Preliminary feasibility report. Diagn Ther Endosc 2011:7 pp Article ID 182435
5. Figueiredo IN, Moreno JC, Prasath VBS, Figueiredo PN (2012) A segmentation model and application to endoscopic images. In: Campilho A, Kamel M (eds) International conference on image analysis and recognition (ICIAR 2012). Springer LNCS, vol 7325. Aveiro, Portugal, pp 164–171 (June 2012)
6. Prasath VBS, Pelapur R, Palaniappan K (2014) Multi-scale directional vesselness stamping based segmentation for polyps from wireless capsule endoscopy. Figshare (June 2014)
7. Karargyris A, Bourbakis N (2010) A survey on wireless capsule endoscopy and endoscopic imaging. a survey on various methodologies presented. IEEE Eng Med Biol Mag 29(1):72–83
8. Asari KV (2000) A fast and accurate segmentation technique for the extraction of gastrointestinal lumen from endoscopic images. Med Eng Phys 22(2):89–96
9. Asari KV, Srikanthan T (2002) Segmenting endoscopic images using adaptive progressive thresholding: a hardware perspective. J Syst Architect 47(9):759–761
10. Chan TF, Vese LA (2001) Active contours without edges. IEEE Trans Image Process 10 (2):266–277
11. Prasath VBS, Figueiredo IN, Figueiredo PN, Palaniappan K (2012) Mucosal region detection and 3D reconstruction in wireless capsule endoscopy videos using active contours. In: 34th IEEE/EMBS international conference, San Diego, USA, pp 4014–4017 (September 2012)
12. Mumford D, Shah J (1989) Optimal approximations by piecewise smooth functions and associated variational problems. Commun Pure Appl Math 42(5):577–685
13. Osher S, Sethian JA (1988) Fronts propagating with curvature-dependent speed: algorithms based on Hamilton-Jacobi formulations. J Comput Phys 79(1):12–49
14. Prasath VBS (2009) Color image segmentation based on vectorial multiscale diffusion with inter-scale linking. In: Chaudhury S, Mitra S, Murthy CA, Sastry PS, Sankar K.P (eds) Third international conference on pattern recognition and machine intelligence (PReMI-09). Springer LNCS, vol 5909. Delhi, India, pp 339–344 (December 2009)
15. Moreno JC, Prasath VBS, Proenca H, Palaniappan K (2014) Brain MRI segmentation with fast and globally convex multiphase active contours. Comput Vis Image Underst 125:237–250
16. Prasath VBS, Pelapur R, Palaniappan K, Seetharaman G (2014) Feature fusion and label propagation for textured object video segmentation. In: SPIE Defense + Security (DSS). Baltimore, MD, USA (May 2014) In Geospatial Info Fusion and Video Analytics, IV
17. Comaniciu D, Meer P (2002) Mean shift: A robust approach toward feature space analysis. IEEE Trans Pattern Anal Mach Intell 24(5):603–619
18. Sandberg B, Chan TF, Vese L (2002) A level-set and Gabor-based active contour algorithm for segmenting textured images. Technical Report, pp 02–39, UCLA CAM (2002)
19. Bresson X, Esedoglu S, Vandergheynst P, Thiran J, Osher S (2007) Fast global minimization of the active contour/snake model. J Math Imaging Vis 28(2):151–167

Effect of Feature Selection on Kinase Classification Models

Priyanka Purkayastha, Akhila Rallapalli, N.L. Bhanu Murthy,
Aruna Malapati, Perumal Yogeeswari and Dharmarajan Sriram

Abstract Classification of kinases will provide comparison of related human kinases and insights into kinases functions and evolution. Several algorithms exist for classification and most of them failed to classify when the dimension of feature set large. Selecting the relevant features for classification is significant for variety of reasons like simplification of performance, computational efficiency, and feature interpretability. Generally, feature selection techniques are employed in such cases. However, there has been a limited study on feature selection techniques for classification of biological data. This work tries to determine the impact of feature selection algorithms on classification of kinases. We have used forward greedy feature selection algorithm along with random forest classification algorithm. The performance was evaluated by selecting the feature subset which maximizes Area Under the ROC Curve (AUC). The method identifies the feature subset from the datasets which contains the physiochemical properties of kinases like amino acid, dipeptide, and pseudo amino acid composition. An improvised performance of classification is noted for feature subset than with all the features. Thus, our method indicates that groups of kinases are classifiable with maximum AUC, if good subsets of features are used.

P. Purkayastha (✉) · A. Rallapalli · N.L. Bhanu Murthy · A. Malapati · P. Yogeeswari ·
D. Sriram
BITS Pilani Hyderabad Campus, Shameerpet, RR District, Hyderabad 500078, AP, India
e-mail: p2011001@hyderabad.bits-pilani.ac.in

A. Rallapalli
e-mail: f2011223@hyderabad.bits-pilani.ac.in

N.L. Bhanu Murthy
e-mail: bhanu@hyderabad.bits-pilani.ac.in

A. Malapati
e-mail: arunam@hyderabad.bits-pilani.ac.in

P. Yogeeswari
e-mail: pyogee@hyderabad.bits-pilani.ac.in

D. Sriram
e-mail: dsriram@hyderabad.bits-pilani.ac.in

© The Author(s) 2015
N.B. Muppalaneni and V.K. Gunjan (eds.), *Computational Intelligence in Medical Informatics*, Forensic and Medical Bioinformatics,
DOI 10.1007/978-981-287-260-9_8

81

Keywords Forward greedy feature selection · Random forest · Area under the ROC curve · Kinases classification

1 Introduction

Kinases have become the most studied class for drug target and play important role in metabolism, cell signaling, cellular transport, protein regulation, secretory processes, and many other cellular pathways [1]. Hence, kinases annotations are necessary for understanding the pathway related to signal transduction and for understanding the disease pathway associated with impairment of kinases activity [2]. So, classification of kinases will provide comparison of related human kinases and insights into kinases functions and evolution. Currently the PROT R package in R was to extract the features for kinases. With a wide range of features, choosing an accurate features subset for classification is not an easy task. Feature selection techniques are employed in such cases. Feature selection method identifies subset of features, based on which a classifier will be trained. Feature selection is an important step in training a classifier with a subset of features, instead of training the classifier with the entire features of a dataset [3, 4].

Feature selection is beneficial in various ways. First, a packed subset of features can alleviate the curse of dimensionality and alienates the overfitting problem which is usually encountered during training a classifier. Secondly, a model performs with a highly accurate extent by removing the noisy features and preprocessing the dataset. Thirdly, the accurate and essential subset of the dataset can be used with the significantly reduced computational cost. Finally, an illustrative subset of feature can make the model output more understandable and reasonable. Computational cost can be reduced by randomly eliminating a number of features. The major challenge lies in search of feature subset that leads to enhanced performance of a classifier by removing the redundant and unnecessary features. Consequently, the efficacy of a feature selection method is commonly assessed by the performance of the final model trained with the feature subset [4].

Researchers working on this problem have explored building classification model using influential features, in which classification accuracy has thought to be the best measure for assessing the performance of a classifier. However, it has been pointed out that accuracy is not always a suitable assessment metric and the Area Under the ROC Curve (AUC) has been proven as a better performance metric in evaluation with classification accuracy [5]. A classifier with minimum cost is more required than a classifier with high accuracy [6]. In this work, we have used forward greedy feature selection algorithm along with random forest classification algorithm and evaluated the performance based on the feature subset which maximizes AUC.

Table 1 The number of sequences belongs to each of the kinases family

Kinases families	Number of sequences
AGC	63
Atypical	43
CAMK	79
CK1	11
CMGC	63
OTHERS	80
RGC	4
STE	47
TK	89
TKL	42
Total	530

2 Materials and Methods

The kinases groups are majorly classified into 10 groups, as shown in Table 1. The sequences are collected from Manning et al. classification and grouped accordingly [7]. The dataset consist of 530 sequences, which includes physiochemical properties like amino acid composition, dipeptide composition, and pseudo amino acid composition of human kinases obtained from Prot R package.

3 Results and Discussion

The performance of the random forest classifier was evaluated for all the 10 group of kinases namely, (i) Tyrosine Kinases (TK); (ii) Tyrosine Kinase-Like (TKL); (iii) Protein kinases with the families with A, G, and C (AGC); (iv) Calmodulin/Calcium regulated kinases (CAMK); (v) Casein kinase 1 (CK1); (vi) Cyclin-dependent kinases (CDKs), mitogen-activated protein kinases (MAP kinases), glycogen synthase kinases (GSK), and CDK-like kinases (CMGC); (vii) Other kinases; (viii) STE kinases; (ix) RGC kinases; and (x) Atypical kinases.

The classification of kinases to all the 10 groups was done using the amino acid, dipeptide, and pseudo amino acid composition. The amino acid composition is of fixed pattern and is of 20 amino acids. The dipeptide composition consists of 400 amino acid dipeptide combinations. The pseudo amino acid composition also contains detail information about three properties, hydrophobicity, hydrophilicity, and side chain mass. The pseudo amino acid composition contains features of length 80. Amino acid and dipeptide composition was used by Bhasin and Raghava for nuclear receptor classification [8]. Earlier pseudo amino acid composition was used Krajewski and Tkacz for structural classification of proteins based on pseudo amino acid composition with a measure of accuracy [9]. But recently, both theoretical and empirical studies revealed that a classifier with the highest accuracy extent might not be idyllic in real-world problems. Instead, the AUC has been demonstrated as the alternative approach and measure to evaluate the performance of any classifier. Therefore, we attempt to develop classification models using Random Forest classifier [10]. We have developed an algorithm by building the model using 2/3rd of the training dataset and remaining 1/3rd of the test dataset. The test datasets were partitioned randomly.

The performance of the classifier was evaluated using AUC. Suppose we need to select k feature subset from a feature set of $F = \{f_1, f_2, ..., f_m\}$. Forward greedy search builds model by considering one feature at a time and by calculating AUC for each of them [11]. Then, the combination feature subsets are ranked based on the descending order. The combination of features with maximum AUC value is selected for classification of kinases.

The performance of the subset of features from the amino acid, dipeptide, and pseudo amino acid composition are shown in Figs. 1, 2 and 3 respectively. As shown in Figs. 1, 2 and 3, an improvised performance of the classifier was obtained after applying forward greedy feature selection algorithm than with all the features. The feature selection maximizes AUC measures for all the 10 classes of kinases.

AUC measured for amino acid, dipeptide and pseudo amino acid composition for all 10 classes shows a major difference in AUC measure with all 20 features than compared with feature selected subsets. The AUC measure was seen for amino acid composition in Fig. 1. The major difference in AUC measure was found for RGC class than compared with other kinase classes. The feature subset with respect to RGC class was found to be reduced to three features instead of all 20 features, using forward greedy algorithm. A negligible difference was identified in case of CK1 and TK classes, which could further be studied for marking a significant difference in all the kinase classes.

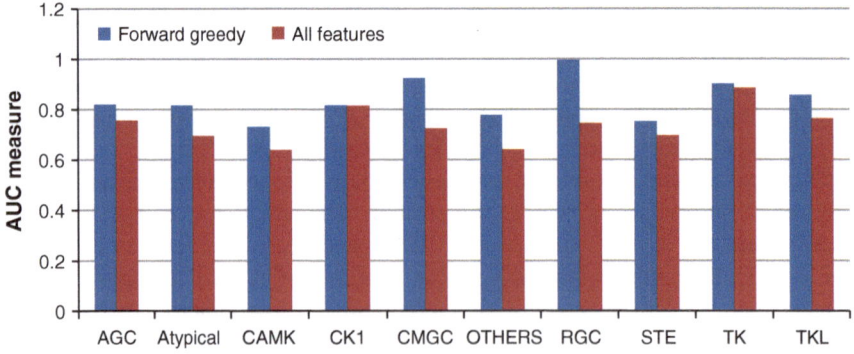

Fig. 1 The performance of the classifier for all 10 kinase classes using amino acid composition using feature subset and all features

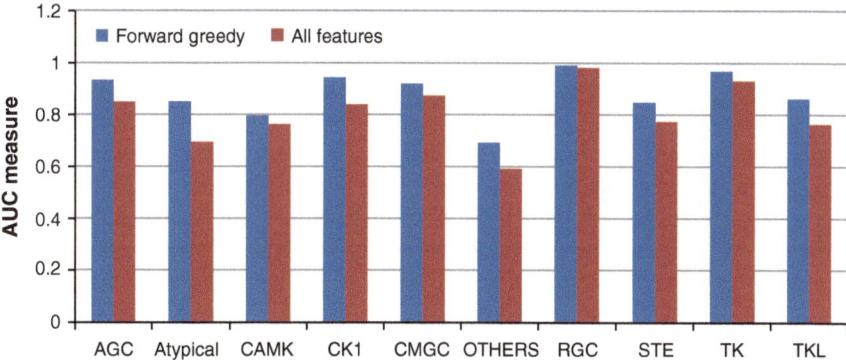

Fig. 2 The performance of the classifier for all 10 kinase classes using dipeptide composition using feature subset and all features

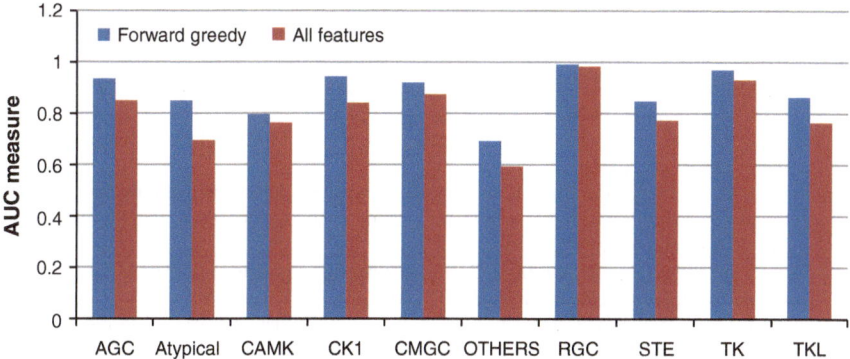

Fig. 3 The performance of the classifier for all 10 kinase classes using pseudo amino acid composition using feature subset and all features

Similarly, the performance in AUC measure was compared using all 400 features and subset of features using dipeptide amino acid composition as shown in Fig. 2. The major difference was found in case of atypical class of kinases using all 400 features and subset of 4 features (using forward greedy) and the negligible difference was found in case of TK class. Similarly, for pseudo amino acid composition the difference measured was found to be more in case of atypical than compared with all other classes and very less in case of RGC as shown in Fig. 3. The number of features generated using forward greedy was found to contain six features with highest AUC measure. This brings us to the hypothesis that kinases can be classified with maximum AUC extent, if good subsets of features are used.

4 Conclusion

In this paper, we have shown the pros of feature selection method for identifying the feature subset for classification of kinases. The performance of the classification model is shown using the feature subset and using all the features. The evaluation of the performance was done by measuring AUC. The random forest classifier is able to classify kinase groups with a better AUC measure for feature subsets than compared with all the features. But the difference in AUC measure was found to be negligible for a few classes of kinase like RGC class using amino acid composition, atypical class using dipeptide, and pseudo amino acid composition which indicates that group of kinases are classifiable with maximum AUC extent, if a good subset of features are used. Further, feature selection method could useful to classify large set of biological data and for dimensionality reduction.

References

1. Cohen P (2002) Protein kinases–the major drug targets of the twenty-first century? Nat Rev Drug Discov 1(4):309–315
2. Zhang J, Yang PL, Gray NS (2009) Targeting cancer with small molecule kinase inhibitors. Nat Rev Cancer 1(9):28–39
3. Ding C, Peng H (2003) Minimum redundancy feature selection from microarray gene expression data. In: Proceedings of the IEEE computer society conference on bioinformatics, pp 523–528. Washington, DC
4. Tang K, Suganthan P, Yao X (2006) Gene selection algorithms for microarray data based on least squares support vector machine. BMC Bioinform 7:95
5. Huang J, Ling CX (2005) Using AUC and accuracy in evaluating learning algorithms. IEEE Trans Knowl Data Eng 17(3):299–310
6. Rui W, Tang K (2009) Feature selection for maximizing the area under the ROC curve. In: Data mining workshops, 2009. ICDMW'09. IEEE international conference on. IEEE
7. Manning G et al (2002) The protein kinase complement of the human genome. Science 298 (5600):1912–1934
8. Bhasin M, Raghava GP (2004) Classification of nuclear receptors based on amino acid composition and dipeptide composition. J Biol Chem 279(22):23262–23266
9. Krajewski Z, Tkacz E (2013) Protein structural classification based on pseudo amino acid composition using SVM classifier. Biocybern Biomed Eng 33(2):77–87
10. Breiman Leo (2001) Random forests. Mach Learn 45(1):5–32
11. Bradley Andrew P (1997) The use of the area under the ROC curve in the evaluation of machine learning algorithms. Pattern Recogn 30(7):1145–1159

Rheumatoid Arthritis Candidate Genes Identification by Investigating Core and Periphery Interaction Structures

Sachidanand Singh, V.P. Snijesh and J. Jannet Vennila

Abstract Rheumatoid arthritis (RA) is a long-term systemic inflammatory disease that primarily attacks synovial joints and ultimately leads to their destruction. The disease is characterized by series of processes such as inflammation in the joints, synovial hyperplasia, and cartilage destruction leading to bone erosion. Since RA being a chronic inflammatory complex disease, there is a constant need to develop novel and dynamic treatment to cure the disease. In the present research, network biology and gene expression profiling technology are integrated to predict novel key regulatory molecules, biological pathways, and functional network associated with RA. The microarray datasets of synovial fibroblast (SF) (GSE7669) and macrophages (GSE10500 and GSE8286), which are the primary cells in the synovium and reported as the key players in the pathophysiology of RA, were considered for identification of signature molecules related to RA. The statistical analysis was performed using false discovery rate (FDR), t-test, one-way anova, and Pearson correlation with favorable p-value. The K-core analysis depicted the change in network topology which consisted of up- and downregulated genes network, resulted in six novel meaningful networks with seed genes OAS2, VCAN, CPB1, ZNF516, ACP2, and OLFML2B. Hence, we propose that, differential gene expression network studies will be a standard step to elucidate novel expressed gene(s) globally.

Keywords Rheumatoid arthritis · Gene expression profiling · Synovial fibroblast · Macrophages · K-core analysis

S. Singh (✉) · V.P. Snijesh
Department of Bioinformatics, School of Biotechnology and Health Sciences,
Karunya University, Coimbatore, India
e-mail: sachidanand@karunya.edu

J. Jannet Vennila
Department of Biotechnology, School of Biotechnology and Health Sciences,
Karunya University, Coimbatore, India

© The Author(s) 2015 87
N.B. Muppalaneni and V.K. Gunjan (eds.), *Computational Intelligence in Medical Informatics*, Forensic and Medical Bioinformatics,
DOI 10.1007/978-981-287-260-9_9

1 Introduction

Network biology and gene expression profiling with microarrays have turned into a typical approach for finding genes and biological pathways which are correlated with diverse complex diseases [3, 5, 12]. Biological network analysis combines coexpressed gene networks to predict novel associations and key regulatory molecules. It integrates independent data as well as biologically relevant information which emerge as a reliable method to find out meaningful novel biological network(s) [4, 24, 39].

Rheumatoid arthritis (RA) is a systemic autoimmune disease characterized by inflammation, synovial hyperplasia, cartilage destruction, and bone erosion [31]. The involvement of immune cells and inflammatory molecules are basic hallmark of RA [13]. The cells of synovium are basically divided into synovial fibroblast (SF) and macrophages. The primary function of SF and macrophages are secretion of hyaluronic acid and phagocytosis, respectively [1, 23]. They are involved in production of various inflammatory cytokines and chemokines, which in turn attract more inflammatory molecules to the synovium [16, 35]. As together they also initiate in production of Matrix Metalloproteinase (MMP), Vascular Endothelial Growth Factor (VEGF), formation of ectopic germinal layers, and over expression of Major Histocompatibility Complex Class II (MHC II) [6, 15, 18]. Recent studies have reported that both SF and macrophages play crucial role in progressive joint inflammation and destruction as they are found larger in pannus and inflamed synovial membrane than in normal joints [2]. Current research work focuses on identification of key regulatory molecules from synovial cell layers.

Microarray technology has been implemented to identify gene expression level which can be further used to detect transcriptionally altered key signature molecules involved in the pathophysiology of RA. At current times, various studies have made to predict differentially expressed genes (DEGs) in RA using multiple gene expressions [17]. In the current study, molecular expression profiles of human macrophages and SF from Gene Expression Omnibus (GEO) were analyzed to identify promising genes for RA. Our investigation provided a valuable methodology to analyze these novel genes which are involved in the pathophysiology of RA.

2 Materials and Methods

2.1 Data Collection

The research work focused on the key regulatory molecules involved in the pathophysiology of RA. Synovial lining consists of two types of cells SF and macrophages which are reported as the key players in the pathogenesis to RA [25, 38]. RA related microarray datasets of SF and macrophages were retrieved from GEO (http://www.ncbi.nlm.nih.gov/geo/). Overall 3 datasets, 2 from synovial macrophage

(GSE10500 and GSE8286) and 1 from SF (GSE7669) under Affymetrix platform were selected. The series GSE10500 consisted of 8 samples with 5 from RA patients and 3 from normal donors [44]. The dataset GSE8286 was associated with differentiation of monocyte to macrophage which consisted of total of 9 samples. Three each samples from monocyte at 0, 16, and 168 h were collected [22]. GSE7669 comprised of 6 samples each from RA patients and osteoarthritis patients [10].

2.2 Data Preprocessing

Each dataset was incorporated into GeneSpring GX 12.6 and raw signal values of each probe sets were normalized by baseline to median of control samples using Robust Multiarray Average (RMA) algorithm. It normalizes the raw signals by creating expression matrix from Affymetrix data by involving background correction of raw signals, log2 transformation, and quantile normalization (Fig. 1) [14]. All samples in the datasets were experimentally grouped based on the content of the sample. 8 samples of GSE10500 were grouped as 3 control (normal donors) and 5 diseased (RA patients). Similar grouping was made in the case of GSE7669. But for the dataset GSE8286 grouping was made based on time and condition. Differentiation of monocyte to macrophage under 0–16 h and 0–168 h conditions were grouped (Fig. 2).

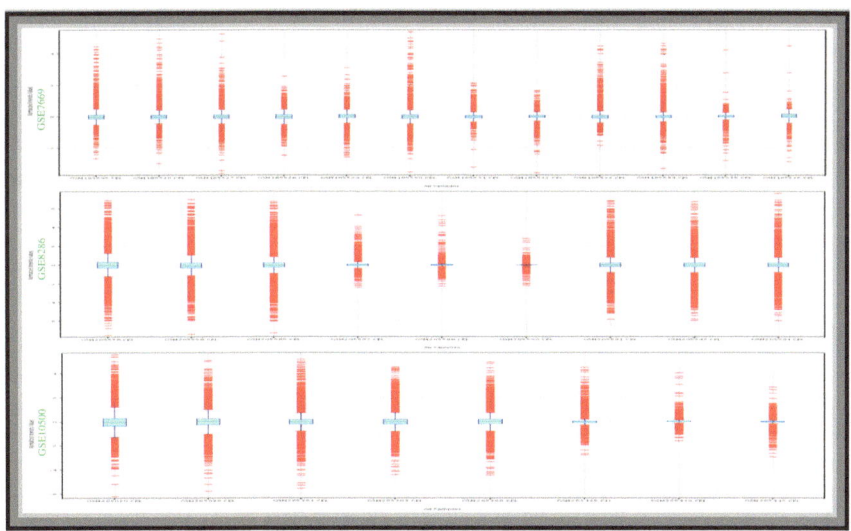

Fig. 1 Box plot of normalized values for series GSE7669, GSE8286, and GSE10500

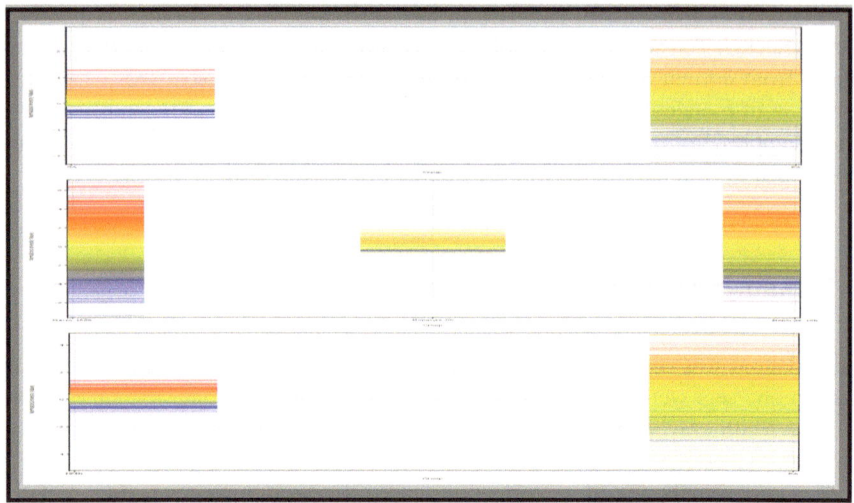

Fig. 2 Profile plot of group for series GSE7669, GSE8286, and GSE10500

2.3 Differentially Expressed Genes

The collected datasets differ in terms of number of samples, experimental conditions, and type. To identify the DEGs, statistical analysis was carried out based on one-way annova, t-test, and Benjamini–Hochberg False Discovery Rate (FDR). Paired t-test and unpaired t-test were performed for datasets based on the number of sample on grouping. Datasets having equal number of controlled and treated samples were performed with paired t-test. Datasets GSE10500, GSE8286, and GSE7669 followed t-test unpaired, one-way anova, and t-test paired, respectively. Direct Benjamini–Hochberg FDR with p-value cutoff 0.05 and Fold Change (FC) value ≥ 1.5 and ≤ -1.5 were carried out for entire datasets to identify the significant expressed genes in the dataset.

2.4 Protein Network Analysis

Gene–gene correlation map, based on Pearson correlation with centroid linkage clustering method, was constructed for both upregulated and downregulated genes. The relationship between entire datasets was developed using Pearson's correlation coefficient (R-value) [27]. Symmetrical $n \times n$ matrix was generated at R-value cutoff 0.9 and above. From the matrix, genes with high correlation value were imported to Cytoscape [34] for the network construction of coexpressed genes.

In any biological network, node represents gene or protein and the interactions between these nodes are termed as degree which represents various pathways through which they are associated [36]. Basically, biological networks are scale free networks with statistically and functionally significant interacting molecules [33]. Cytoscape plugins-Network analyzer and M-Code were implemented for identifying the topology of the network keeping into consideration of K-core for identifying promising hubs [30]. It was further validated by biological analysis by Biointerpreter (http://www.biointerpreter.com/biointerpreterv3/) [42].

3 Result and Discussion

3.1 Expression Characteristics for RA Datasets

Differentially expressed genes were studied using statistical measures for each series of macrophages and fibroblast of synovial membrane. RMA algorithm normalized the raw signals of all the series which were under Affymetrix platform [9]. The multiple raw signals were compared using multiple hypothesis testing, p-value (≤ 0.05), and FC = 1.5 (≥ 1.5 and ≤ -1.5). The series GSE7669 of fibroblast resulted in 79 DEGs with 35 up- and 54 downregulated genes. There were 691 DEGs in the dataset of GSE10500 with 424 and 267 up- and downregulated genes, respectively. The series GSE8286 in multiple groups were reported a total of 145 DEGs in which group 0–16 h was in 3 and 1 and group 0–168 h in 51 and 90 up- and downregulated genes, respectively (Table 1).

Higher correlation coefficient ($r \geq 0.9$) with Pearson algorithm implied in gene–gene correlation of DEGs resulted in 403 genes interactome (edges = 81,140) for downregulated and 510 genes interactome (edges = 183,064) for upregulated genes. The whole interactome gave a good landscape of highly correlated coexpression matrix.

Table 1 List of up- and downregulated genes in different series of SF and Macrophages

Fibroblast_GSE7669			
Upregulated		Downregulated	
35		54	
Macrophage_GSE10500			
Upregulated		Downregulated	
424		267	
Macrophage_GSE8286			
0–16 h		0–168 h	
Upregulated	Downregulated	Upregulated	Downregulated
3	1	51	90

3.2 K-Core Hubs Identification for DEGs

Both the up- and downregulated networks ended up with three clusters (Rank 1, Rank 2, and Rank 3) with K-core of 15 (Table 2). The seed nodes obtained through the analysis were ZNF516, ACP2, and OLFML2B for downregulated (Fig. 3) whereas OAS2, VCAN, and CPB1 for upregulated networks (Fig. 4). Seed nodes are the highest scoring nodes in the cluster which connect all other nodes and represent the central node of connectivity [21]. The M-CODE score value ranged from 67 to 846 for Rank 1–3 (Table 2).

Table 2 K-core analysis with highly clustered seed genes for down- and upregulated networks

Clusters	Node	Score	Edges	Seed
K-core for downregulated genes (15–20)				
Rank 1	265	528	69,984	ZNF516
Rank 2	86	177	7,309	ACP2
Rank 3	49	103	2,351	OLFML2B
K-core for upregulated genes (15–20)				
Rank 1	424	846	179,358	OAS2
Rank 2	86	96	2,440	VCAN
Rank 3	49	67	991	CPB1

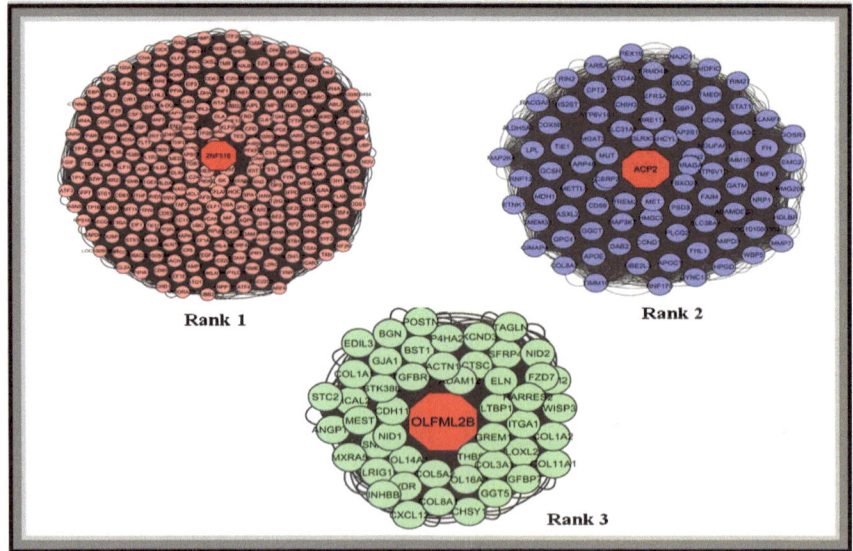

Fig. 3 Seed genes on K-core analysis for downregulated network

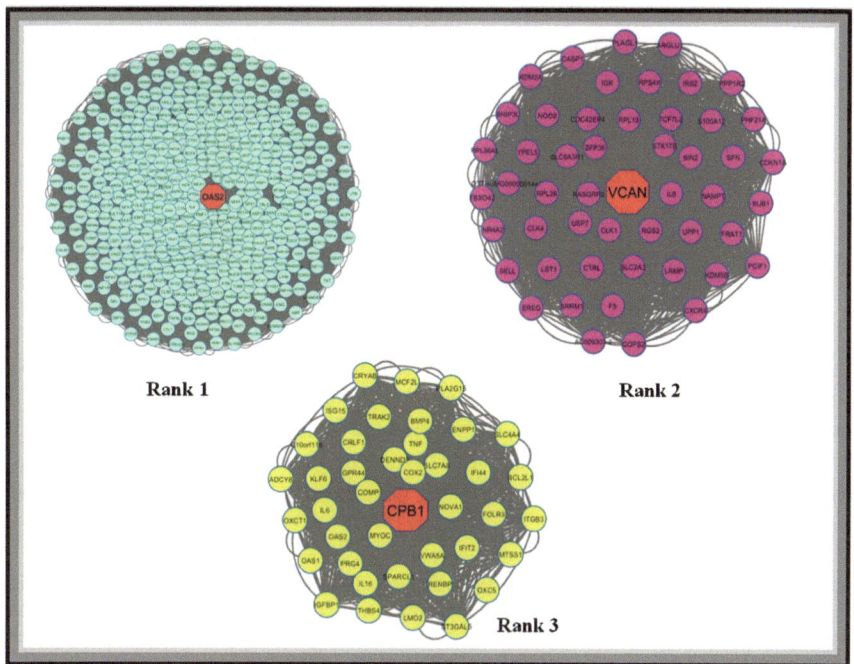

Fig. 4 Seed genes on K-core analysis for upregulated network

Common biological properties like angiogenensis, antigen processing and presentation, immune response, vasculogenesis, chemotaxis, and inflammation confirmed that these nodes may play role in pathophysiology of RA from the cytoplasmic and nuclear level. 40 % of genes identified under different categories of synovial membrane (Rank 1–3 for up and Rank 1–3 for down) pointed to have correlation with the disease hyperlipidemia which is a genetic disorder of increased blood fats causing cardiovascular risk which is further connected with RA as it doubles the risk in RA patient [11, 26]. All the ranks from the both networks showed their influence in cytokine–cytokine interaction, TGF beta signaling, ECM receptor interaction, Osteoclast differentiation, and Wnt receptor signaling pathways. Under upregulated networks, the seed gene Chondroitin Sulfate Proteoglycan (VCAN) showed its involvement in glycosaminoglycan and hyaluronic acid binding [29, 43]. Being a major component of cartilage, it is the main point of attraction for CD44 for initiating the inflammation in synovial cells [8]. Oligoadenylate Synthetase 2 (OAS2) and Carboxypeptidase B1 (CPB1) showed crossregulation and downregulation with autoimmune diseases, respectively [37]. OAS2 was involved in IFN beta signaling pathway whereas CPB1which is present in plasma helps in fibrin clot by acting as a procoagulant [20, 32, 40]. CPB1 also initiates osteopontin, C5a, and bradykinin-like proinflammatory molecules [7]. Under downregulated networks, Zinc Finger Protein 516 (ZNF516), Acid phosphatase 2, lysosomal (ACP2), and Olfactomedin-like 2B (OLFML2B) showed their specific response in stress activity specifically in immune response [19, 28]. These genes have already indicated upregulation in cardiovascular and hormone regulation. ACP2 and OLFML2B also showed their response to hyperlipidemia disease which was connected to RA [41]. All the proteins molecules reported through network analysis showed their effect in pathophysiology of RA. Still there is wide scope to further investigate these networks which can explore more genes/proteins related to RA.

4 Conclusion

The proposed research work is an expression network study frame work for RA synovial cells which may play significant role in its pathophysiology. The outcome resulted in six candidate seed genes which gave a better understanding of the progression of RA by different pathways involvement. This part of information may lead to better understand and manage RA in future by considering these potential targets in therapeutics.

Acknowledgments This research work was supported by Science Engineering and Research Board-Department of Science and Technology, New Delhi and Karunya University, Coimbatore, Tamil Nadu.

References

1. Anitua E, Sánchez M, Nurden AT, Zalduendo MM, De La Fuente M, Azofra J, Andía I (2007) Platelet-released growth factors enhance the secretion of hyaluronic acid and induce hepatocyte growth factor production by synovial fibroblasts from arthritic patients. Rheumatology 46(12):1769–1772
2. Athanasou NA (1995) Synovial macrophages. Ann Rheum Dis 54(5):392
3. Bauer JW, Bilgic H, Baechler EC (2009) Gene-expression profiling in rheumatic disease: tools and therapeutic potential. Nat Rev Rheumatol 5(5):257–265
4. Begley TJ, Rosenbach AS, Ideker T, Samson LD (2002) Damage recovery pathways in Saccharomyces cerevisiae revealed by genomic phenotyping and interactome mapping1 1 NIH grants RO1-CA-55042 and P30-ES02109; NIH training grant ES07155 and National Research Service Award F32-ES11733 (to TJB). Mol Cancer Res 1(2):103–112
5. Chand Y, Alam MA (2012) Network biology approach for identifying key regulatory genes by expression based study of breast cancer. Bioinformation 8(23):1132
6. Cho CH, Koh YJ, Han J, Sung HK, Lee HJ, Morisada T, Koh GY (2007). Angiogenic role of LYVE-1-positive macrophages in adipose tissue. Circ Res 100(4):e47–e57
7. Du XY, Zabel BA, Myles T, Allen SJ, Handel TM, Lee PP, Leung LL (2009) Regulation of chemerin bioactivity by plasma carboxypeptidase N, carboxypeptidase B (activated thrombin-activable fibrinolysis inhibitor), and platelets. J Biol Chem 284(2):751–758
8. Fujimoto T, Kawashima H, Tanaka T, Hirose M, Toyama-Sorimachi N, Matsuzawa Y, Miyasaka M (2001) CD44 binds a chondroitin sulfate proteoglycan, aggrecan. Int Immunol 13 (3):359–366

9. Gautier L, Cope L, Bolstad BM, Irizarry RA (2004) Affy—analysis of Affymetrix GeneChip data at the probe level. Bioinformatics 20(3):307–315

10. Geurts J, Vermeij EA, Pohlers D, Arntz OJ, Kinne RW, van den Berg WB, van de Loo FA (2011) A novel Saa3-promoter reporter distinguishes inflammatory subtypes in experimental arthritis and human synovial fibroblasts. Ann Rheum Dis 70(7):1311–1319

11. Han C, Robinson DW, Hackett MV, Paramore LC, Fraeman KH, Bala MV (2006) Cardiovascular disease and risk factors in patients with rheumatoid arthritis, psoriatic arthritis, and ankylosing spondylitis. J Rheumatol 33(11):2167–2172

12. Häupl T, Stuhlmüller B, Grützkau A, Radbruch A, Burmester GR (2010) Does gene expression analysis inform us in rheumatoid arthritis? Ann Rheum Dis 69(suppl 1):i37–i42

13. Huber LC, Distler O, Tarner I, Gay RE, Gay S, Pap T (2006) Synovial fibroblasts: key players in rheumatoid arthritis. Rheumatology 45(6):669–675

14. Irizarry RA, Hobbs B, Collin F, Beazer-Barclay YD, Antonellis KJ, Scherf U, Speed TP (2003) Exploration, normalization, and summaries of high density oligonucleotide array probe level data. Biostatistics 4(2):249–264

15. Ito TK, Ishii G, Chiba H, Ochiai A (2007) The VEGF angiogenic switch of fibroblasts is regulated by MMP-7 from cancer cells. Oncogene 26(51):7194–7203

16. Jones JA, Chang DT, Meyerson H, Colton E, Kwon IK, Matsuda T, Anderson JM (2007) Proteomic analysis and quantification of cytokines and chemokines from biomaterial surface-adherent macrophages and foreign body giant cells. J Biomed Mater Res Part A 83 (3):585–596

17. Kim TH, Choi SJ, Lee YH, Song GG, Ji JD (2014) Gene expression profile predicting the response to anti-TNF treatment in patients with rheumatoid arthritis; analysis of GEO datasets. Jt Bone Spine 81(4):325–330

18. Kobayashi H, Puri P, O'Briain DS, Surana R, Miyano T (1997) Hepatic overexpression of MHC class II antigens and macrophage-associated antigens (CD68) in patients with biliary atresia of poor prognosis. J Pediatr Surg 32(4):590–593

19. Kovar DR, Wu JQ, Pollard TD (2005) Profilin-mediated competition between capping protein and formin Cdc12p during cytokinesis in fission yeast. Mol Biol Cell 16(5):2313–2324

20. Lepus CM, Song JJ, Wang Q, Wagner CA, Lindstrom TM, Chu CR, Robinson WH (2014) Brief report: carboxypeptidase B serves as a protective mediator in osteoarthritis. Arthritis Rheumatol 66(1):101–106

21. Li X, Wu M, Kwoh CK, Ng SK (2010) Computational approaches for detecting protein complexes from protein interaction networks: a survey. BMC Genom 11(suppl 1):S3

22. Liu H, Shi B, Huang CC, Eksarko P, Pope RM (2008) Transcriptional diversity during monocyte to macrophage differentiation. Immunol Lett 117(1):70–80

23. Mantovani A, Sozzani S, Locati M, Allavena P, Sica A (2002) Macrophage polarization: tumor-associated macrophages as a paradigm for polarized M2 mononuclear phagocytes. Trends Immunol 23(11):549–555

24. Marcotte EM, Pellegrini M, Thompson MJ, Yeates TO, Eisenberg D (1999) A combined algorithm for genome-wide prediction of protein function. Nature 402(6757):83–86

25. Mor A, Abramson SB, Pillinger MH (2005) The fibroblast-like synovial cell in rheumatoid arthritis: a key player in inflammation and joint destruction. Clin Immunol 115(2):118–128

26. Nelson RH (2013) Hyperlipidemia as a risk factor for cardiovascular disease. Prim Care Clin Off Pract 40(1):195–211

27. Ng A, Bursteinas B, Gao Q, Mollison E, Zvelebil M (2006) pSTIING: a 'systems' approach towards integrating signalling pathways, interaction and transcriptional regulatory networks in inflammation and cancer. Nucleic Acids Res 34(suppl 1):D527–D534

28. Pantoja-Uceda D, Arolas JL, García P, López-Hernández E, Padró D, Aviles FX, Blanco FJ (2008) The NMR structure and dynamics of the two-domain tick carboxypeptidase inhibitor reveal flexibility in its free form and stiffness upon binding to human carboxypeptidase b†‡. Biochemistry 47(27):7066–7078

29. Perides G, Biviano F, Bignami A (1991) Interaction of a brain extracellular matrix protein with hyaluronic acid. Biochim Biophys Acta (BBA) (General Subjects) 1075(3):248–258

30. Raman MP, Singh S, Devi PR, Velmurugan D (2012) Uncovering potential drug targets for tuberculosis using protein networks. Bioinformation 8(9):403
31. Rossol M, Schubert K, Meusch U, Schulz A, Biedermann B, Grosche J, Wagner U (2013) Tumor necrosis factor receptor type I expression of CD4+ T cells in rheumatoid arthritis enables them to follow tumor necrosis factor gradients into the rheumatoid synovium. Arthritis Rheum 65(6):1468–1476
32. Sadler AJ, Williams BR (2008) Interferon-inducible antiviral effectors. Nat Rev Immunol 8 (7):559–568
33. Sengupta U, Ukil S, Dimitrova N, Agrawal S (2009) Expression-based network biology identifies alteration in key regulatory pathways of type 2 diabetes and associated risk/ complications. PLoS ONE 4(12):e8100
34. Shannon P, Markiel A, Ozier O, Baliga NS, Wang JT, Ramage D, Ideker T (2003) Cytoscape: a software environment for integrated models of biomolecular interaction networks. Genome Res 13(11):2498–2504
35. Smith RS, Smith TJ, Blieden TM, Phipps RP (1997) Fibroblasts as sentinel cells. Synthesis of chemokines and regulation of inflammation. Am J Pathol 151(2):317
36. Snijesh VP, Singh S (2014) Molecular modeling and network based approach in explaining the medicinal properties of nyctanthes arbortristis, lippia nodiflora for rheumatoid arthritis. J Bioinform Intell Control 3(1):31–38
37. Song JJ, Hwang I, Cho KH, Garcia MA, Kim AJ, Wang TH, Robinson WH (2011) Plasma carboxypeptidase B downregulates inflammatory responses in autoimmune arthritis. J Clin Invest 121(9):3517–3527
38. Szekanecz Z, Koch AE (2007) Macrophages and their products in rheumatoid arthritis. Curr Opin Rheumatol 19(3):289–295
39. Tornow S, Mewes HW (2003) Functional modules by relating protein interaction networks and gene expression. Nucleic Acids Res 31(21):6283–6289
40. van Baarsen LG, Wijbrandts CA, Rustenburg F, Cantaert T, van der Pouw TC (2010) Regulation of IFN response gene activity during infliximab treatment in rheumatoid arthritis is associated with clinical response to treatment. Arthritis Res Ther 12(1):11
41. van Oorschot RA, Birmingham V, Porter PA, Kammerer CM, VandeBerg JL (1993) Linkage between complement components 6 and 7 and glutamic pyruvate transaminase in the marsupialMonodelphis domestica. Biochem Genet 31(5–6):215–222
42. Varatharajan S, Karathedath S, Velayudhan SR, Srivastava A, Mathews V, Balasubramanian P (2013) Harnessing gene expression profiling in search of new candidate genes for Ara-C resistance in acute myeloid leukemia. Blood 122(21):1299
43. Westling J, Gottschall P, Thompson V, Cockburn A, Perides G, Zimmermann D, Sandy J (2004) ADAMTS4 (aggrecanase-1) cleaves human brain versican V2 at Glu405-Gln406 to generate glial hyaluronate binding protein. Biochem J 377:787–795
44. Yarilina A, Park-Min KH, Antoniv T, Hu X, Ivashkiv LB (2008) TNF activates an IRF1-dependent autocrine loop leading to sustained expression of chemokines and STAT1-dependent type I interferon–response genes. Nat Immunol 9(4):378–387